BIOLOGICAL AND MEDICAL PHYSICS
BIOMEDICAL ENGINEERING

Springer
New York
Berlin
Heidelberg
Hong Kong
London
Milan
Paris
Tokyo

BIOLOGICAL AND MEDICAL PHYSICS BIOMEDICAL ENGINEERING

The fields of biological and medical physics and biomedical engineering are broad, multidisciplinary and dynamic. They lie at the crossroads of frontier research in physics, biology, chemistry, and medicine. The Biological & Medical Physics/Biomedical Engineering Series is intended to be comprehensive, covering a broad range of topics important to the study of the physical, chemical and biological sciences. Its goal is to provide scientists and engineers with textbooks, monographs, and reference works to address the growing need for information.

Continued on page 133 after Index

Howard C. Berg

E. coli in Motion

With 42 Figures, 1 in Full Color

Howard C. Berg
Department of Molecular and Cellular Biology
Harvard University
Cambridge, MA 02138
USA
hberg@biosun.harvard.edu

Library of Congress Cataloging-in-Publication Data
Berg, Howard C., 1934–
    E. coli in motion / Howard C. Berg.
        p.   ; cm.—(Biological and medical physics biomedical engineering)
    Includes bibliographical references and index.
    ISBN 0-387-00888-8 (hc. : alk. paper)
    1. Escherichia coli.   2. Microorganisms–Motility.   I. Title.   II. Series.
    [DNLM: 1. Escherichia coli—pathogenicity. QW 138.5.E8 B493e 2003]
    QR82.E6B47 2003
    579.3'42—dc21   2003045491

ISBN 0-387-00888-8   Printed on acid-free paper.

Printed in the United States of America.

9 8 7 6 5 4 3 2 1   SPIN 10922125

www.springer-ny.com

Springer-Verlag   New York Berlin Heidelberg
A member of BertelsmannSpringer Science+Business Media GmbH

Series Preface

The fields of biological and medical physics and biomedical engineering are broad, multidisciplinary and dynamic. They lie at the crossroads of frontier research in physics, biology, chemistry, and medicine. The Biological & Medical Physics/Biomedical Engineering Series is intended to be comprehensive, covering a broad range of topics important to the study of the physical, chemical and biological sciences. Its goal is to provide scientists and engineers with textbooks, monographs, and reference works to address the growing need for information.

Books in the series emphasize established and emergent areas of science including molecular, membrane, and mathematical biophysics; photosynthetic energy harvesting and conversion; information processing; physical principles of genetics; sensory communications; automata networks, neural networks, and cellular automata. Equally important will be coverage of applied aspects of biological and medical physics and biomedical engineering such as molecular electronic components and devices, biosensors, medicine, imaging, physical principles of renewable energy production, advanced prostheses, and environmental control and engineering.

Elias Greenbaum
Oak Ridge, TN

Preface

Most bacteria are small, about one micrometer in diameter: ten thousand cells laid out side by side span the width of one's finger. Nevertheless, many species are motile. They swim using propellers (called flagella) that extend out into the external medium or, in the case of spirochetes, that rotate within the cell envelope. One marine bacterium appears to use submicroscopic external oars. Other common bacteria, equipped with large numbers of flagella, swarm rapidly over surfaces. Some bacteria glide over surfaces by extending and retracting thin filaments (called pili) that stick to the substratum at their distal ends, a kind of fly casting. Others move particles linked to the substratum along their outer membranes, by a mechanism as yet unknown. Bacteria of all kinds respond to changes in their environment, for example, to changes in temperature, light intensity, or chemical composition. In short, they move in a purposeful manner.

I have been interested in this world for more than 30 years. When I began, more was known about the genetics and biochemistry of the bacterium *Escherichia coli* than of any other free-living thing. So that has been the organism of choice. The emphasis has been on the responses of this organism to chemical stimuli: chemotaxis. Early work on the motile behavior of bacteria had been done with larger species, more easily seen in the light microscope, so these also are of interest.

How, exactly, does *E. coli* behave? What is the machinery that makes this behavior possible? How is the construction of this machinery programmed? How does this machinery work? And finally, what remains to be discovered?

Since *E. coli* is microscopic and lives in an aqueous environment, the physical constraints that it has had to master are very different from those that we encounter. For example, *E. coli* knows nothing about inertia, only about viscous drag: it cannot coast. It knows nothing about transport by bulk flow, only about diffusion;

as we will see, it can go where the grass is greener, but it has to wait for its dinner. So the methods that its cells use to move and sample their environment are strange to us. This is part of *E. coli's* charm.

This book is designed for the scientist or engineer, not trained in microbiology, who would like to learn more about living machines. However, it also should be accessible to the educated layman and of interest to the expert. I try to build on first principles. However, if you are overwhelmed by the facts that appear in a given chapter, please read on: the figures might suffice. References are given as entrée to the literature and a tribute to those who have done the work.

My own research has been supported by the Research Corporation, the U.S. National Science Foundation, the U.S. National Institutes of Health, and the Rowland Institute for Science. Much of the writing was done while a Fellow of the John Simon Guggenheim Foundation. Space for thought was provided by the Lorentz Institute, Leiden.

A large number of capable people have contributed to the body of knowledge to be described here: molecular geneticists, biochemists, microbial physiologists, physicists. Some, no doubt, will disagree with my emphasis. I can claim only a small part of this work as my own, built on the labor of students, postdocs, and other colleagues. The real hero is *E. coli.* If nothing else, I hope that this book will convince you that *E. coli* demands our admiration and respect.

HOWARD C. BERG

Cambridge, Massachusetts
August 2003

Contents

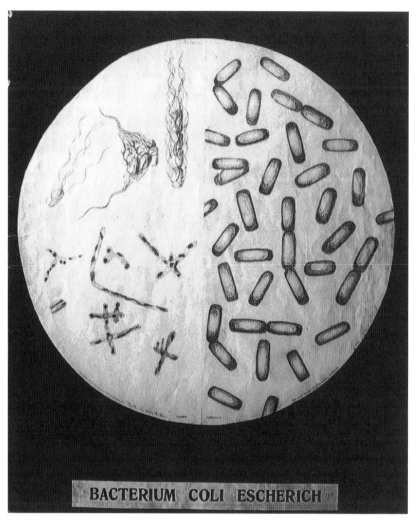

BACTERIUM COLI ESCHERICH

E. coli, circa 1900. A color poster used in lectures by Martinus W. Beijerinck, founder of the Delft School of Microbiology, drawn by his sister Henriëtte. Courtesy of Lesley A. Robertson.

1
Why *E. coli*?

Heritage

E. coli (*Escherichia coli*) is a bacterium that lives in your gut. It is one of the simplest and best understood living things, yet remarkably sophisticated. Fossil remains of bacteria are found in rocks that are billions of years old. Thus, creatures of *E. coli*'s kind are a thousand times older than we are. Yet we are closely related. *E. coli* stores genetic information in the same way that we do, reads that information in the same way, and synthesizes the same kinds of molecular tools for carrying out basic cellular functions. Many of the enzymes (catalytic proteins) designed for harvesting energy or crafting molecular building blocks have nearly identical structures. Thus, a number of early solutions to life's problems found by bacteria have been passed down to us.

Size and Shape

E. coli is very small. Its cells are rod-shaped, about 2.5 micrometers (μm) long—10,000 end to end span 1 inch—by about 0.8 μm in diameter, with hemispherical end caps. Imagine a microscopic cocktail sausage. As the cell grows, it gets longer and then divides in the middle. In a warm, rich nutrient broth, this takes only 20 minutes. The cell has a thin three-layered wall enclosing a relatively homogeneous molecular soup, called the cytoplasm. It does not have a nucleus, other membrane-enclosed organelles, or any cytoskeletal elements (rope-like or rod-like components) typical of higher cells, such as those composing the human body. However, some elements of this kind are built into the cell wall. And *E. coli* does have external organelles, thin straight filaments, called pili, that enable it to attach to specific substrata, and thicker, longer helical filaments, called flagella, that enable it to swim.

Habitat

E. coli lives a life of luxury in the lower intestines of warm-blooded animals, including humans. Once expelled, it lives a life of penury and hazard in water, sediment, and soil. *E. coli* is a minor constituent of the human gut. A typical stool contains as many as 10^{11} (100 billion) bacteria per cubic centimeter (cm^3). Up to 10^9 (1 billion) of these are *E. coli*. The majority of the other bacteria are strictly anaerobic, and thus unable to live in the presence of oxygen outside of the body. Cells of *E. coli* can live with or without oxygen, and thus survive (with luck) until they find another host. The particular species that we are going to learn about is called K-12. It lives in the laboratory. It was isolated in 1922 from the feces of a convalescent diphtheria patient and maintained at Stanford University, beginning in 1925, in a departmental culture collection. It has been in captivity for so long that it is no longer able to colonize the human gut. Fed well, however, it grows to a density similar to that of its siblings there, to some 10^9 cells per cm^3: the population of India in a spoonful.

Pathogenicity

Most but not all *E. coli* are friendly. But some cause urinary tract infections. Others cause diarrheal diseases and contribute to infant mortality. The latter strains carry islands of DNA not present in cells that normally inhabit the human gut. An exceptionally nasty one, called O157:H7—these are names for particular types of cell-surface (O) and flagellar antigens (H)—can cause severe or fatal renal or neurological complications. It also carries genes encoding toxins acquired from a relative, *Shigella*, a dysentery bacillus. But these are the exceptions. Common strains such as K-12 are our friends. Among other things, they help prevent invasion of the gut by yeast and fungi, organisms that are far less benign.

Preeminence

E. coli was first identified in the intestinal flora of infants by the German pediatrician Theodor Escherich (1885), who called it *Bacterium coli commune*. It was named for Escherich in 1920. For a review of Escherich's work, see Bettelheim (1986). *E. coli* was a useful organism for studies of bacterial physiology, because it was readily accessible, generally benign, and grew readily on chemically defined media. Thus, it came to be used for dissection of biochemical pathways; for studies of bacterial viruses, of bacterial and viral genetics, of the regulation of gene expression, of the nature of the genetic code, of gene replication, and of protein synthesis; and, in the present age of genetic engineering, for the manufacture of proteins of commercial value.

Motile Behavior

E. coli also has been a model organism for the study of the molecular biology of behavior, the primary focus of this book. *E. coli* swims. It modifies the way in which it swims to move toward regions in its environment that it deems more favorable. Each flagellum is driven at its base by a reversible rotary motor, driven by a proton flux. The cell's ability to migrate in a particular direction results from the control of the direction of rotation of these flagella. This control is effected by intracellular signals generated by receptors in the cell wall that count molecules of interest that impinge on the cell surface. What were the chemical and physical constraints that *E. coli* had to meet to devise such mechanisms? How does all of this work? Indeed, what is it like being a microscopic organism living in an aqueous environment? Can we understand *E. coli*'s world? The answers to these questions are fascinating, in part, because that world is so very different from our own. We will try to come to grips with the life of this distant yet intimate relative, and step, as it were, into *E. coli*'s shoes.

This is a cross-field endeavor. Physicists seek precise descriptions of how cells move and of the kinds of measurements that they make on their environment. Biochemists are interested in the structures and interactions of molecules that monitor the external environment, pass information from the outside to the inside of the cell, process sensory data, and effect a response. Geneticists identify the genes that specify these molecules and learn how these genes are turned on and off. With *E. coli*, the physics, biochemistry, and genetics are all readily accessible.

Simplicity

In grappling with such basic questions, one has to start somewhere, preferably with something simple. Compare the sizes and dates of origin of various nervous systems outlined in Fig. 1.1. We are astonishingly complex. The neocortex in humans is a multilayered sheet of cells, about 1 millimeter (mm) thick, almost large enough to cover your desk top. Each cubic millimeter contains about 40,000 cells, each making some 20,000 connections. The axons that make the connections in each cubic millimeter total several kilometers (km) long! A pyramidal cell in the olfactory cortex of the cat is shown in Fig. 1.2. *E. coli* is about the size of one synaptic spine, one of the little bumps seen on the enlarged views of the dendrites.

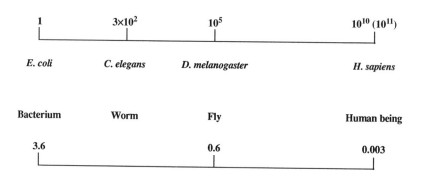

FIGURE 1.1. Approximate number of cells in different nervous systems, shown on a logarithmic scale. The estimate for humans is low. The species that are best understood are *E. coli, Caenorhabditis elegans, Drosophila melanogaster,* and *Homosapiens sapiens,* respectively. The scale is also historic: bacteria go back a few billion years, worms and flies a few hundred million, and humans only a few million.

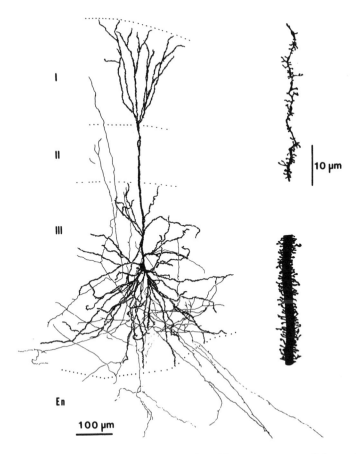

FIGURE 1.2. One pyramidal cell from the olfactory cortex of the cat. The main body of the cell is at the center, the dendrites (branched extensions that receive synaptic inputs) are at the top, and the axon (a filamentous extension that sends signals to other cells) is at the bottom. Segments of dendrites are shown enlarged at the right, covered with synaptic spines, each about the size of *E. coli*. (Tseng & Haberly, 1989, Fig. 2, reprinted with permission.)

To understand *E. coli*'s behavior, we will touch on a number of other topics of interest in cellular and molecular biology. The total extent of this knowledge is vast. For *E. coli*, the "bible" is a book called *Escherichia coli and Salmonella: Cellular and Molecular Biology* (Neidhardt et al., 1996). This is a two-volume work put together by 237 authors with the help of 10 editors comprising 155 chapters: 3008 pages in 8½ × 11-inch format. There is more information there than can be fathomed by any one human brain.

Genes and Behavior

When one starts with *E. coli*, debates about the importance of genes for behavior (Weiner, 1999) have an air of unreality. Of course genes play an essential role in behavior: for *E. coli*, that is all there is. The function of the product of essentially every behavioral gene is clear. There is no evidence that *E. coli* knows anything about associative learning, yet its behavior is remarkably sophisticated. One can even approach the question of free will. *E. coli's* behavior is fundamentally stochastic: cells either run or tumble. Their motors spin either counterclockwise or clockwise. Transitions between the latter states are thermally activated. *E. coli's* irritability derives from the basic laws of statistical mechanics. This irritability is modulated by the cell's reading of its environment. Our thoughts might be triggered in the same way, by changes in states that are thermally activated.

References

Bettelheim, K. A. 1986. Commemoration of the publication 100 years ago of the papers by Dr. Th. Escherich in which are described for the first time the organisms that bear his name. *Zentralbl. Bakteriol. Mikrobiol. Hyg. [A]* 261:255–265.

Escherich, T. 1885. Die Darmbakterien des Neugeborenen und Säuglings. I. *Fortschr. Med.* 3:515–522.

Neidhardt, F. C., R. I. Curtiss, J. L. Ingraham, et al., editors. 1996. *Escherichia coli* and *Salmonella*: Cellular and Molecular Biology. ASM Press, Washington, DC.

Tseng, G.-F., and L. B. Haberly. 1989. Deep neurons in piriform cortex. I. Morphology and synaptically evoked responses including a unique high-amplitude paired shock facilitation. *J. Neurophysiol.* 62:369–385.

Weiner, J. 1999. Time, Love, Memory: A Great Biologist and His Quest for the Origins of Behavior. Vintage, New York.

2
Larger Organisms

Seventeenth Century

Antony van Leeuwenhoek was the first person to see bacteria, and it was their motion that captured his attention. However, he was not the first person to use a microscope or to describe cells. But the single-lens instruments that he made himself had fewer aberrations than the compound instruments of his day, so he was able to see more. And his curiosity was insatiable. His work on little animals—he called them animalcules—is available to the modern reader through translations from the archaic Dutch by the British microbiologist Clifford Dobell (1932), who published them on the 300th anniversary of van Leeuwenhoek's birth. Van Leeuwenhoek described what he saw in letters written in ink, still jet black, sealed with red wax, and sent from Delft to London, to Henry Oldenberg, the secretary of the Royal Society. Oldenberg translated bits and pieces and published them in the *Transactions of the Royal Society*. My favorite is the 18th letter in which van Leeuwenhoek describes animalcules in water from his well. He was curious about the effect that pepper might have, so he ground up some in a blue porcelain pot and mixed it in. A number of larger animalcules died out, until on 6 August 1676 he saw large numbers of very small ones:

I now saw very plainly that these were little eels, or worms, lying all huddled up together and wriggling; just as if you saw, with the naked eye, a whole tubful of very little eels and water, with the eels a-squirming among one another: and the whole water seemed to be alive with these multifarious animalcules. This was for me, among all the marvels that I have discovered in nature, the most marvellous of all; and I must say, for my part, that no more pleasant sight has ever yet come before my eye than these many thousands of living creatures, seen all alive in a little drop of water, moving among one another, each several creature having its own proper motion.

Dobell believed these to be cells of a large spiral organism, *Spirillum volutans*, shown at the bottom of Fig. 2.1. Van Leeuwenhoek's first drawings of bacteria came later, in 1683, when he examined the scruff on his teeth. He never saw bacterial flagella, but he marveled at their evident small size. In his 26th letter of 1678 he wrote,

but I see, besides these, other living animalcules which are yet more than a hundred times less, and on which I can make out no paws, though from their structure and the motion of their body I am persuaded that they too are furnished with paws withal: and if their paws be proportioned to their body, like those of the bigger creatures, upon which I can see the paws, then, taking their measure at but a hundred times less, it follows that a million of their paws together make up but the thickness of a hair of my beard; while these paws, besides their organs for motion, must also be furnished with vessels whereby nourishment must pass through them.

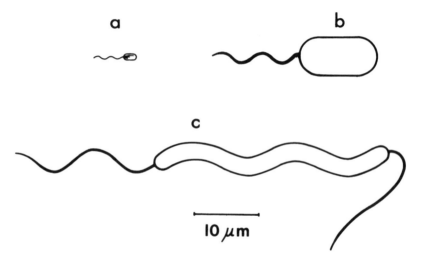

FIGURE 2.1. Scale drawings of some flagellated bacteria whose behavior has been studied. (a) *E. coli*. About four filaments arise at random from the sides of the cell and form a bundle that appears near one pole. The bundle pushes the cell. When one or more filaments transiently changes its direction of rotation, the cell alters course. (b) *Chromatium okenii*. About 40 filaments arise at one pole. The bundle either pushes or pulls the cell. When the filaments change their direction of rotation, the cell backs up. (c) *Spirillum volutans*, shown swimming from left to right. The body is helical. About 25 filaments arise at each pole. Those on the left are in the tail configuration; those on the right in the head configuration. When the filaments in either bundle change their directions of rotation and flip over (tail to head, head to tail), the cell swims in the opposite direction, as if reflected in a mirror.

Like other early naturalists, van Leeuwenhoek believed that microorganisms were macroorganisms writ small!

Very few of van Leeuwenhoek's microscopes survive, but they have been studied by modern methods and found to pass muster. There is no doubt that he was able to see what he claimed to see (Ford, 1985, 1991). Robert Brown (1828) used a single-lens instrument in his studies of brownian motion, of which we will hear more later.

Van Leeuwenhoek was born in the same year and christened in the same church as Johannes Vermeer. Vermeer died in his early 40s and van Leeuwenhoek was the executor of his estate. But van Leeuwenhoek lived on to his early 90s. He was buried in the Oude Kerk at Delft, where his daughter Maria erected a monument in his memory.

Nineteenth Century

Bacterial flagella were first seen in 1836, when the German naturalist Christian Ehrenberg (1838) found an enormous bacterium in the brook below the church of Ziegenhayn, near Jena. This organism is shown at the upper right in Fig. 2.1. He called it *Monas okenii* (now *Chromatium okenii*), in honor of Oken, the founder of the society then meeting in Jena. *C. okenii* is a photosynthetic red sulfur bacterium. It converts hydrogen sulfide (H_2S) to elementary sulfur (S), which appears as granules in the cell cytoplasm. Ehrenberg took these granules to be stomach cells. He also identified the ovary!

In 1883, Theodor Engelmann, a German physiologist working in Utrecht, found in the waters of a branch of the Rhine a similar organism, which he called *Bacterium photometricum*. He was amazed by its behavior toward light. If, while looking through the microscope at a population of swimming cells, he passed his hand between the light source and the specimen stage, every cell backed up. This gave Engelmann the impression of fright, so he called it a shock reaction. He then showed that cells accumulate in a spot of light, not because they like the light, but because they are afraid of the dark. They swim into the spot perfectly well, but are not able to get out. When crossing a dark-light boundary from dark to light, they tend to speed up; when crossing in the opposite direction, they back up. Engelmann observed a similar response to carbon dioxide: when cells in a hanging drop were suddenly exposed to this gas, they also backed up.

Engelmann (1881a,b) studied the responses of a variety of other bacteria to oxygen. Some species swam toward higher concentrations of oxygen and others toward lower concentrations. Some were more discriminating, liking some oxygen but not too much. Species of the first kind, when placed in an aqueous suspension under a square coverglass, accumulated at the edge of the coverglass. Species of the second kind fled to the middle. Species of the third kind formed a square array a certain distance away from the edge of the coverglass. When Engelmann blew hydrogen gas at this preparation, the square got larger; when he used oxygen, the square got smaller. Then he had a bright idea: use cells that like oxygen as analytical chemists to indicate where oxygen is generated in a green plant during photosynthesis. The answer proved to be the chloroplast. Engelmann reviewed these experiments in 1894. Figure 2.2 shows one of his drawings of an agal cell called *Spirogyra* that has a spiral chloroplast. When the cell was illuminated with a spot of light, as shown on the left, the bacteria accu-

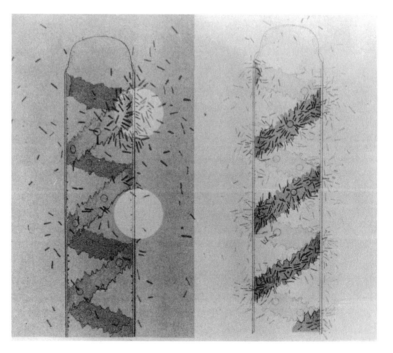

FIGURE 2.2. Part of a drawing by Engelmann of oxygen-loving bacteria (possibly *Bacillus subtilis*) responding to the illumination of *Spirogyra*, an algal cell with a spiral chloroplast. The original figure is in color.

mulated near its surface, but only if the spot impinged on the chloroplast. If the cell was illuminated uniformly, the bacteria accumulated in a spiral array.

Engelmann had an interesting career. He spent 30 years as a physiologist in Utrecht. He also was a cellist, and his second wife, Emma, a pianist. They were friends of Brahms: to wit, the Engelmann Quartet (Kamen, 1986). He ended his days in Berlin. He was best known for work on excitation of muscle in the heart.

Some beautiful sketches of *Chromatium okenii* were published by Manabu Miyoshi (1898). While soaking in the Yumoto Spa near Nikko, he became interested in tufts of sulfur in water from the volcanic springs. In ditches and pools nearby he found reddish blooms that proved to be nearly pure cultures of *C. okenii*. His drawings of their responses to gradients of chemical attractants or repellents are shown in Fig. 2.3. Miyoshi's reddish blooms are still there. I visited the Yumoto Spa in the spring of 1999 with Chi Aizawa, a colleague then at Teikyo University. We took samples back to his laboratory in Utsunomiya, put them under the microscope, and passed our hands between the light source and the specimen stage. The shock reaction is alive and well.

The capillary assay used by Miyoshi had been developed in the 1880s by Wilhelm Pfeffer (1884), a botanist working in Tübingen. He used it to study the responses of a number of different species of bacteria to a variety of chemicals. Initially, he thought that bacteria could steer toward the mouth of a capillary tube containing a chemical attractant. So he coined the term "*chemotaxis*." As we shall see later, bacteria are not able to steer. *E. coli* uses a strategy called "klinokinesis with adaptation." But molecular biologists are not fond of nomenclature, so the term *chemotaxis* has stuck. Pfeffer's capillary assay was perfected by Julius Adler, a biochemist at the University of Wisconsin in Madison, who began the modern work on chemotaxis in *E. coli*. We will hear about this shortly. Pfeffer ended his days in Leipzig. He was best known for his work on osmotic pressure (measured with porous porcelain pots impregnated with copper ferrocyanide).

The Golden Age of Microbiology

One does not read about work on the motile behavior of bacteria in books on the history of microbiology. Their emphasis, understandably, is on bacteria that cause disease. Robert Koch isolated

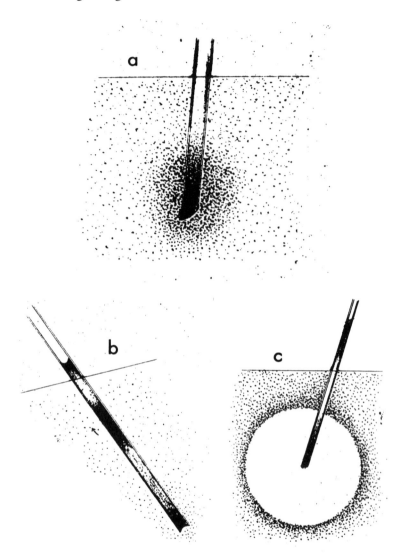

FIGURE 2.3. Drawings by Miyoshi of the responses of *C. okenii* to a chemical attractant (ammonium nitrate, 0.3% w/v) or a chemical repellent (malic acid, 0.5% w/v) diffusing from the tip of a capillary tube. In the response to the attractant, the bacteria accumulate near the mouth of the capillary tube and then swim inside. The original figures are in color.

the anthrax bacillus in 1876, the tubercle bacillus in 1882, and the cholera vibrio in 1883. He published his postulates specifying the criteria for proof of the cause of infectious disease in 1884. Beginning in 1880, Louis Pasteur demonstrated immunization by attenuated bacteria (or virus) for cholera in birds, anthrax in sheep, erysipelas in pigs, and rabies in dogs. The use of agar as a bacteriological medium appeared in 1882, the Gram stain in 1884, the Petri plate in 1887, and the Institut Pasteur in 1888. This was the golden age of medical microbiology. So why bother about bacterial motility? One exception was Koch's (1877) photographs of stained flagella that appeared as part of a discourse on technical methods for bacterial examination.

Some pathogenic organisms are motile and others are not, as shown in Table 2.1. The importance of motility and chemotaxis

TABLE 2.1. Motility of the main bacterial pathogens discovered between 1877 and 1906.

Pathogen, disease	Bacterial genus or species
Cells motile	
Typhoid fever	*Salmonella typhi*
Cholera	*Vibrio cholerae*
Tetanus	*Clostridium tetani*
Diarrhea	*Escherichia coli*
Food poisoning	*Salmonella enteritidis*
Botulism	*Clostridium botulinum*
Paratyphoid	*Salmonella paratyphi*
Syphilis	*Treponema pallidum*
Cells nonmotile	
Anthrax	*Bacillus anthracis*
Suppuration	*Staphylococcus*
Gonorrhea	*Neisseria gonorrhoeae*
Suppuration	*Streptococcus*
Tuberculosis	*Mycobacterium tuberculosis*
Diphtheria	*Corynebacterium diphtheriae*
Pneumonia	*Streptococcus pneumoniae*
Meningitis	*Neisseria meningitidis*
Gas gangrene	*Clostridium perfringens*
Plague	*Yersinia pestis*
Dysentery	*Shigella dysenteriae*
Whooping cough	*Bordetella pertusssis*

Note: The organisms in each category, motile or nonmotile, are listed in the order of discovery. The motile pathogens are all peritrichously flagellated rods (*peri*, around; *trichos*, hair), except *V. cholerae*, which has a single, polar, sheathed flagellum, and *T. pallidum*, which is a spirochete. Some species closely related to the nonmotile pathogens are motile, e.g., *Bacillus cereus*, the *Clostridium* species of the motile group, *Yersinia pseudotuberculosis* and *Bordetella bronchiseptica*. Motile streptococci are rare, but they do exist (members of the group D, or enterococcus, group).

for pathogenicity is still a matter of active study (Ottemann and Miller, 1997), although the common belief is that for a given species, motile cells are more virulent than nonmotile ones. As will be discussed later, chemotaxis offers a cell an enormous advantage for long-range migration, which might be expected to play an important role in invasiveness.

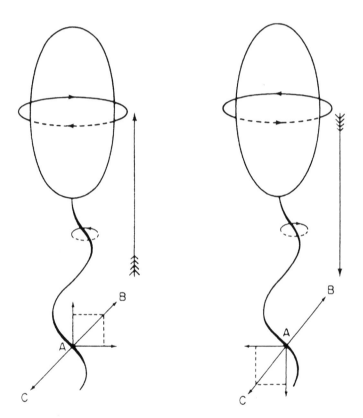

FIGURE 2.4. A sketch of the forward and backward motion of *Chromatium okenii*, according to Buder (1915). During forward motion (left), the flagellar bundle turns rapidly counterclockwise (as seen by an observer behind the cell) and the cell body rolls more slowly clockwise. During backward motion (right) these motions are reversed. Buder analyzed the forces acting on the bundle at an arbitrary point A, assuming (left) that it pressed on the fluid with a force represented by the vector AC and experienced a resistance AB, having components parallel and normal to the helical axis. Scale: the cell body is about 6 μm in diameter.

Early Twentieth Century

The development of dark-field condensers of high numerical aperture enabled Karl Reichert (1909) to determine optimum conditions for visualizing bacterial flagella. An intense cone of light is focussed on the specimen stage in such a way that none directly enters the objective of the microscope. One sees only light scattered from objects in the field of view. Thus, one can visualize such objects even though their dimensions are below the resolving power of the microscope (its ability to distinguish two objects that are next to one another). A single flagellar filament, for example, blooms up to that resolving power (about $0.2\,\mu m$) and looks about 10 times thicker than it actually is. The main problem with the method is that the cell body scatters so much light that it is impossible to distinguish faint objects nearby. However, much of the early work was done with very large organisms, such as *C. okenii* and *S. volutans*, with large flagellar bundles that were easy to see. Figure 2.4 shows an example from the studies of Johannes Buder (1915) on the forward and backward motion of *C. okenii*.

The most extensive work of this kind was done by Peter Metzner (1920) on the behavioral responses of *S. volutans*. The most striking thing about this organism is that reversals of its flagellar bundles are synchronized (unless cells are damaged), even though these bundles are relatively far apart (about $50\,\mu m$). So this organism is capable of rapid long-range intracellular communication, probably electrical. *E. coli* does not have this talent.

Late Twentieth Century

The modern era of work on bacterial behavior began in the 1960s when Julius Adler demonstrated that *E. coli* has a sense of taste, that is, that bacterial chemotaxis is a matter of aesthetics rather than material gain, and when Tetsuo Iino, Sho Asakura, and Goro Eguchi published a series of papers on the assembly of flagellar filaments. The latter work was done with a close relative of *E. coli*, now named *Salmonella enterica* serovar Typhimurium, that I will simply call *Salmonella*. The first picture of *E. coli* published by Adler (an electron micrograph) is shown in Fig. 2.5. For a closer look at the classic literature on bacterial chemotaxis, see Berg (1975).

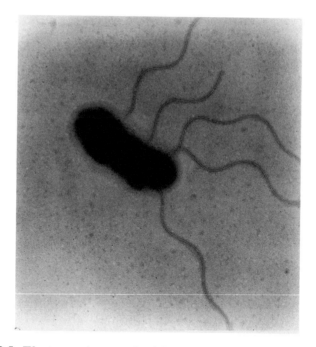

FIGURE 2.5. Electron micrograph of *E. coli* negatively stained (by exposure of a dried sample to the salt of an element of high atomic number, usually tungsten or uranium). Scale: the cell body is about 1 μm in diameter (2 wavelengths of green light). The flagella are extraordinarily thin. (Adler, 1965, Fig. 1, reprinted with permission).

References

Adler, J. 1965. Chemotaxis in *Escherichia coli. Cold Spring Harbor Symp. Quant. Biol.* 30:289–292.

Berg, H. C. 1975. Chemotaxis in bacteria. *Annu. Rev. Biophys. Bioeng.* 4:119–136.

Brown, R. 1828. A Brief Account of Microscopical Observations on the Particles Contained in the Pollen of Plants; and on the General Existence of Active Molecules in Organic and Inorganic Bodies. Richard Taylor, London.

Buder, J. 1915. Zur Kenntnis des *Thiospirillum jenense* und seiner Reaktionen auf Lichtreize. *Jahrb. Wiss. Bot.* 56:529–584.

Dobell, C. 1932. Antony van Leeuwenhoek and His "Little Animals." John Bale, Sons & Danielsson, London. Reprinted by Dover, New York, 1960.

Ehrenberg, C. G. 1838. Die Infusionsthierchen als vollkommene Organismen. Leopold Voss, Leipzig, Germany.

Engelmann, T. W. 1881a. Neue Methode zur Untersuchung der Sauer-
stoffausscheidung pflanzlicher und thierischer Organismen. *Pflügers
Arch. Gesamte Physiol. Menschen Thiere* 25:285–292.

Engelmann, T. W. 1881b. Zur Biologie der Schizomyceten. *Pflügers Arch.
Gesamte Physiol. Menschen Thiere* 26:537–545.

Engelmann, T. W. 1883. *Bacterium photometricum*. Ein Beitrag zur
vergleichenden Physiologie des Licht- und Farbensinnes. *Pflügers
Arch. Gesamte Physiol. Menschen Thiere* 30:95–124.

Engelmann, T. W. 1894. Die Erscheinungsweise der Sauerstoffauss-
cheidung chromophyllhaltiger Zellen im Licht bei Anwendung der
Bacterienmethode. *Pflügers Arch. Gesamte Physiol. Menschen Thiere*
57:375–386.

Ford, B. J. 1985. Single Lens. The Story of the Simple Microscope. Harper
& Row, New York.

Ford, B. J. 1991. The Leeuwenhoek Legacy. Biopress, Bristol, and Farrand
Press, London.

Kamen, M. D. 1986. On creativity of eye and ear: a commentary on the
career of T. W. Engelmann. *Proc. Am. Phil. Soc.* 130:232–246.

Koch, R. 1877. Untersuchungen über Bacterien. VI. Verfahren zur
Untersuchung, zum Conserviren und Photographiren der Bacterien.
Beitr. Biol. Pflanz. 2:399–434.

Metzner, P. 1920. Die Bewegung und Reizbeantwortung der bipolar
begeißelten Spirillen. *Jahrb. Wiss. Bot.* 59:325–412.

Miyoshi, M. 1898. Studien über die Schwefelrasenbildung und die
Schwefelbacterien der Thermen von Yumoto bei Nikko. *J. Coll. Sci.
Imp. Univ. Jap.* 10:143–173.

Ottemann, K. M., and J. F. Miller. 1997. Roles for motility in bacterial-
host interactions. *Mol. Microbiol.* 24:1109–1117.

Pfeffer, W. 1884. Locomotorische Richtungsbewegungen durch chemis-
che Reize. *Unters. Bot. Inst. Tübingen* 1:363–482.

Reichert, K. 1909. Über die Sichtbarmachung der Geisseln und die
Geisselbewegung der Bakterien. *Zentralbl. Bakteriol. Parasitenk.
Infektionskr. Abt. 1 Orig.* 51:14–94.

3
Cell Populations

I will treat the behavior of *E. coli* from the top down, or outside in, beginning with the behavior of cell populations, and then working toward the molecular biology. Imagine an ensemble of self-propelled microscopic particles, moving about in a dilute aqueous medium, robots programmed to respond to external stimuli. How are the robots distributed in space and time?

Chemotactic Rings

A vivid way of demonstrating motile responses of *E. coli* to chemical stimuli (chemotaxis) is to deposit a small drop of a cell suspension on a Petri plate containing semisolid agar (~0.2% w/v) in a nutrient medium. Agar is commonly used at higher concentrations (~2%) as a solid matrix on which to grow discrete bacterial colonies. It is like jello but has the advantage of not being digested by ordinary bacteria. The bacteria grow in this medium, so now the robots are self-replicating. The usual nutrient is tryptone, a mixture of amino acids obtained from a pancreatic digest of casein, a protein found in milk. The structures of three such amino acids are shown in Fig. 3.1. When agar is dilute, motile cells swim through the pores in the gel and spread throughout the plate. They do this in a series of expanding "chemotactic rings," as shown in Fig. 3.2, where clouds of bacteria scatter light and appear white. Adler (1966) found that these rings form as cells consume different nutrients. Wild-type cells, shown at the top, first induce enzymes required for utilization of serine. In front of the outermost ring there is lots of serine; behind it there is practically none. The cells respond to the intervening spatial gradient and move outward. Meanwhile, cells left behind at the point of inoculation induce enzymes required for the utilization of aspartate. In front of the second ring there is lots of aspartate; behind it there is prac-

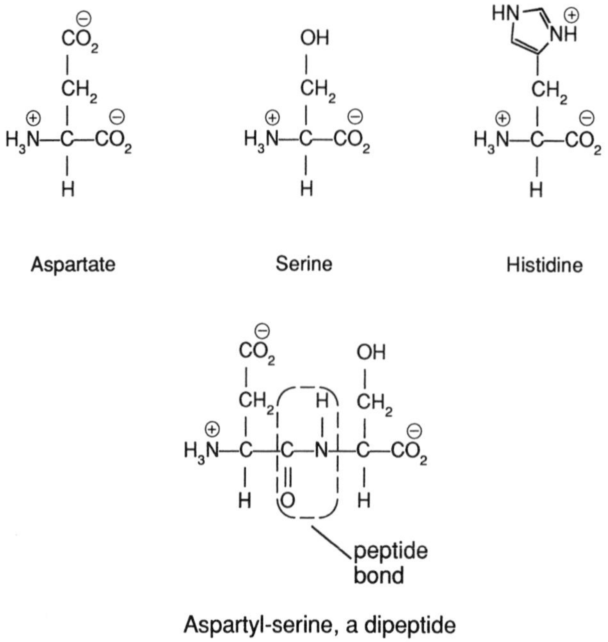

Aspartate   Serine   Histidine

Aspartyl-serine, a dipeptide

FIGURE 3.1. Three amino acids and a dipeptide. In water, the molecules carry positive and negative charges, as shown. Aspartate has a side chain ending in a carboxylic-acid group, serine in a hydroxyl group, and histidine in an imidazole group. The imidazole is a five-membered ring comprising three carbon atoms (indicated by vertices) and two nitrogen atoms. There are 17 other common amino acids that have distinct side chains (not shown). In polypeptides and proteins, amino acids are linked end to end, by the removal of water (H_2O) to form peptide bonds, shown within the dashed line. The atoms shown within the dashed line and the adjacent carbon atoms lie in a plane, and this gives polypeptides favored structural motifs, such as the α-helix and the β-pleated sheet.

tically none. Once again, cells respond to the intervening spatial gradient and move outward. In the course of metabolizing serine and aspartate, the cells deplete most of the oxygen, so next, cells near the bottom of the plate consume threonine anaerobically and move outward in a more diffuse ring. And so on. At the right is a mutant that has lost the ability to taste serine. At the bottom is a mutant that has lost the ability to taste aspartate. At the left is a mutant that swims vigorously but is unable to respond to any attractants or repellents. It fails to form any chemotactic rings. This colony is relatively compact. The cells swim, but they are not

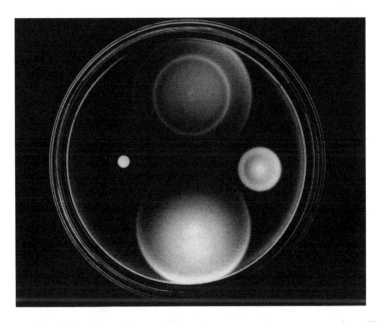

FIGURE 3.2. Behavior of four cell types on a tryptone swarm plate. Top: Wild-type cells, showing chemotactic rings for serine and aspartate. Right: Cells missing the serine receptor. Bottom: Cells missing the aspartate receptor. Left: Smooth-swimming cells unable to process information generated by either chemoreceptor. The plate (8.5 cm dia.) was inoculated in four places by stabbing the agar with a sterile toothpick dipped in a cell suspension and placed in a humid incubator set at 30°C (86°F). About 8 hours later, it was illumined slantwise from below and photographed against a dark background. (Photograph courtesy of J.S. Parkinson, University of Utah.)

able to change directions, so they get trapped in blind alleys in the agar.

Chemotactic rings can be quite sharp, especially if the bacteria metabolize only a single nutrient. A dramatic example is shown in Fig. 3.3, where one inoculum contained cells that could only metabolize the sugar ribose, the other cells that could only metabolize the sugar galactose, with a plate containing a mixture of the two. The cells of either type do not interfere with one another.

Structures of some sugars are shown in Fig. 3.4. These are ring-shaped molecules in which most carbons carry hydroxyl groups. Ribose is a 5-carbon, five-sided ring compound, and galactose is a 6-carbon, six-sided ring compound that differs from glucose only

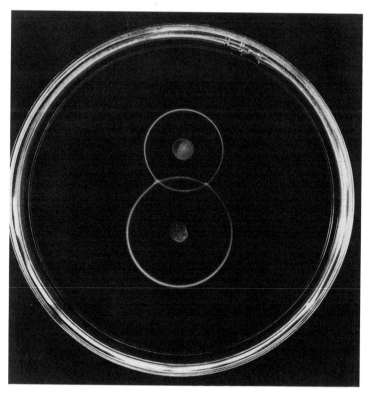

FIGURE 3.3. Behavior of two cell types on a ribose and galactose swarm plate. Both types are chemotactic toward ribose and galactose, but one is unable to metabolize ribose and the other is unable to metabolize galactose. Cells generate a spatial gradient for an attractant only if they consume the attractant. Cells left behind in the original inoculum appear at the center of each ring. (Adler, 1976, and the cover of *Nature*, 26 July 1979, reprinted with permission).

by the position of some of its hydroxyl groups, i.e., whether they are above or below the plane of the ring.

The swarm assay has been enormously useful for finding mutant cells that are defective for chemotaxis or cells that have regained their ability to respond. In the latter case, a single revertant cell appearing at the edge of a compact colony like that of Fig. 3.2 (left) gives rise to a swarm that moves out into regions of the plate otherwise devoid of bacteria. However, the swarm assay is not simply an assay for a behavioral response, because it requires that the cells take up a substrate, and thus generate a chemical gradient, and multiply, to populate the expanding ring. A mutant that

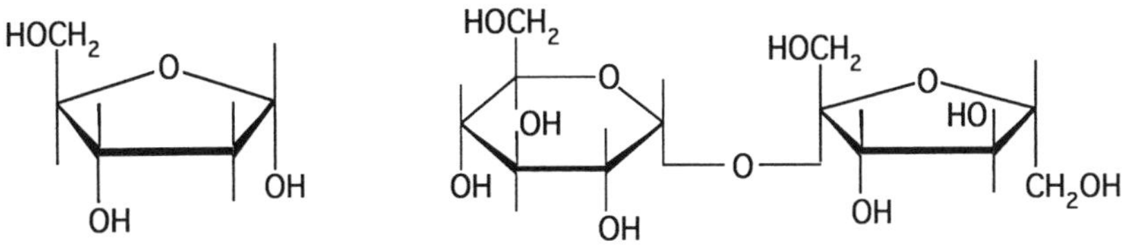

2-Deoxyribose   Sucrose, a disaccharide

FIGURE 3.4. Some sugars. In 2-deoxyribose, the hydroxyl group, present in ribose on carbon 2, is missing. This sugar is part of the backbone of DNA. Sucrose (cane sugar) is a disaccharide: glucose, which has a six-membered ring, is linked to fructose, which has a five-membered ring. The linkage involves removal of water (H_2O). The only difference between fructose and ribose is the placement of the hydroxyl groups. *E. coli* is only weakly chemotactic toward 2-deoxyribose and not chemotactic at all toward sucrose. However, it is chemotactic toward both glucose and fructose. As before, vertices indicate carbon atoms. Hydrogen atoms, located at the end of each unterminated bond, are not shown.

fails to absorb, metabolize, or grow on a substrate will fail to yield a chemotactic ring, even though it might still be able to taste and respond to gradients of that substrate.

Capillary Assay

This led Adler to modernize an assay originally developed by Pfeffer, in which the stimulus is a gradient generated by diffusion of a chemical from the mouth of a capillary tube (Adler, 1969, 1973). But first he needed to find conditions that would support vigorous motility without growth: a chelating agent to protect cells from traces of heavy metals, a buffer to keep the pH between 6 and 7.5 (to keep the acidity close to neutral), and, if the cells were grown aerobically, oxygen to allow utilization of an endogenous energy reserve (Adler and Templeton, 1967). It was found that the presence of glucose or growth above 37°C prevented synthesis of flagella. While Pfeffer looked at the cloud of bacteria that formed near the capillary mouth (Fig. 2.3a), Adler counted the number of bacteria that swam inside (Fig. 2.3b). He did this by making serial dilutions of the contents of the tube, plating aliquots on nutrient agar, and counting colonies (Adler, 1973).

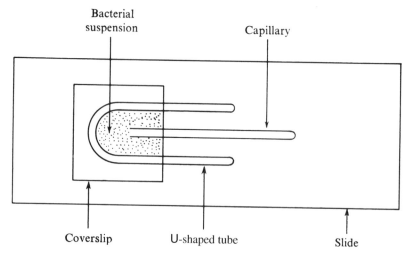

FIGURE 3.5. The apparatus used in Adler's version of the capillary assay. The drawing is to scale (microscope slide 1″ × 3″), except for the width of the capillary tube (200 μm inside diameter). (Adler, 1973, Fig. 1, reprinted with permission.)

A drawing of Adler's version of the capillary assay is shown in Fig. 3.5. A U-shaped spacer made from a glass tube about 1.5 mm in diameter is placed on a glass microscope slide. A coverslip is added as a roof. The space in between is filled with a bacterial suspension. Finally, a capillary tube (200 μm inside diameter) containing a few millimeters of attractant medium is slid along the glass into the pond. It is withdrawn 30 to 60 minutes later.

Chemicals Sensed

Using the capillary assay, Adler was able to show that *E. coli* responds to chemicals that it can neither transport (take up from the surrounding medium) nor metabolize (utilize as a source of energy or raw material). Therefore, the cells must recognize the chemicals per se. Taste will do. Consumption is not necessary. An example is shown in Fig. 3.6.

Some of the different kinds of chemicals to which *E. coli* can respond are listed in Table 3.1. *E. coli* pays attention to things of low molecular weight, among them oxygen, acids and bases, salts, sugars, amino acids, and dipeptides.

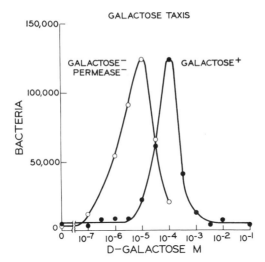

FIGURE 3.6. Numbers of cells entering capillary tubes containing chemicals at different concentrations. Wild-type cells respond strongly to the sugar galactose. Mutant cells also do so, even when defective for uptake and metabolism of galactose. The difference in response is due to the fact that the mutant cells cannot modify the gradient. (Adler, 1987, Fig. 8, reprinted with permission.).

TABLE 3.1. Some chemicals whose gradients strongly affect the motile behavior of wild-type *E. coli.*

Attractants
    Amino acids: e.g., aspartate, serine
    Dipeptides
    Electron acceptors: oxygen, nitrate, fumarate
    Membrane-permeant bases
    Salts at low concentrations
    Sugars and sugar alcohols: e.g., fructose, galacitol, galactose, glucitol, glucose, β-glucosides, maltose, mannitol, mannose, ribose, *N*-acetylglucosamine

Repellents
    Alcohols: e.g., ethanol, isopropanol
    Amino acids: e.g., leucine, isoleucine, valine
    Chemicals at high osmotic strength
    Divalent cations: e.g., cobalt, nickel
    Glycerol or ethylene glycol at high concentrations
    Indole
    Membrane-permeant acids

Other Stimuli

E. coli also is sensitive to changes in temperature, and there is evidence to suggest that cells accumulate in spatial gradients near temperatures at which they were grown (Maeda et al., 1976). Chemoreceptors (e.g., those for aspartate or serine) also serve as temperature sensors, under some conditions responding when the temperature rises and under others when it falls (e.g., Nishiyama et al., 1999).

There are other species of bacteria that respond to light and others that respond to magnetic fields. Most of the former are photosynthetic (use the energy available from light to fix carbon), and they co-opt this machinery to generate behavioral signals. Others have specific photoreceptors (Spudich, 1998; Spudich et al., 2000). Remarkably, chimeric fusions of the latter with chemoreceptors in *E. coli* enable *E. coli* cells to respond to light (Jung et al., 2001). Magnetic bacteria are equipped with arrays of small particles of protein-coated iron oxides (e.g., magnetite) or iron sulfides (e.g., greigite). These cause the cells to line up with the earth's magnetic field, so they behave like swimming compass needles (Blakemore, 1975; Frankel, 1984; Frankel and Blakemore, 1991). *E. coli* (without the photoreceptor transplant) is damaged by light at high intensities; it does not respond to magnetic fields. It is sensitive in the blue, so when working under a microscope at high intensities, one needs to use a cutoff filter that blocks wavelengths shorter than about 500 nm. In the presence of a dye that absorbs light and generates singlet oxygen, cells tumble and then stop swimming (Taylor and Koshland, 1975; Taylor et al., 1979).

More Exotic Patterns

The formation of chemotactic rings (Figs. 3.2 and 3.3) involves interactions between cells that influence one another by removing chemoattractants from the growth medium. Rings also form when chemoattractants are absent in the growth medium, provided that cells excrete a chemoattractant. This can occur when cells are inoculated on soft agar plates containing a nutrient that is readily metabolized aerobically (e.g., an intermediate of the citric-acid cycle, such as fumarate). Under these conditions, the cells excrete aspartate. Wild-type cells, or mutants still able to respond to aspartate, migrate slowly outward in a compact band, metabolizing the nutrient. This band is unstable, because when cells (by growth) reach a critical density, fluctuations in their number, and thus in aspartate concentration, generate gradients steep enough to cause cells to aggregate. This, in turn, increases the local concentration of aspartate. Therefore, starting at a single point and progressing in both directions, the circular band breaks up into a ring of discrete spots. These spots are left behind as cells continue to migrate outward in a compact band. What happens next depends on the concentration of the nutrient. At relatively low concentrations of nutrient, the cells in a spot begin to run out of fuel and stop excreting aspartate. Those that remain motile leave the spot and move outward, rejoining the band. This raises the concentrations of bacteria at the points where they rejoin, and new spots form there. Thus, one gets radial arrays of spots, as shown in Fig. 3.7a. The spots are frozen in place, because the cells, having run out of nutrient, soon stop swimming. At slightly higher

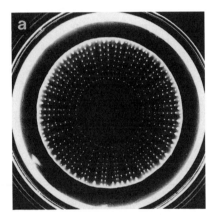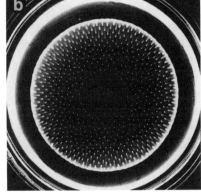

FIGURE 3.7. (a) Cells of a mutant of *E. coli* chemotactic to aspartate but not to serine that have spread outward in a soft-agar plate to form radial arrays of spots. (b) Cells of the same kind that have formed a hexagonal array of spots. The carbon source was α-ketoglutarate (2.5 mM), which is not a chemoattractant. Plate (a) contained, in addition, 2.5 mM hydrogen peroxide, and plate (b) 2.0 mM hydrogen peroxide. The plates were inoculated at the center and incubated for 40 hours at 25°C. They were illuminated slantwise from below and photographed against a dark background. The bright ring near the periphery is an illumination artifact. Other conditions were as described in Budrene and Berg (1991). (Adapted from Berg, 1992, Fig. 1.)

concentrations of nutrient, cells in the spots continue to excrete aspartate, remaining in place until cells in the advancing band increase in number and aggregate anyway. They do so at points midway in between the earlier spots, where the cell densities are higher (because cells were not removed there when the previous set of spots formed). Thus, one gets hexagonal arrays of spots, as shown in Fig. 3.7b. At even higher concentrations of substrate, the cells remain in spots, as before, but tend to move outward as a group. Cells that are not motile are left behind as a streak, so one gets hexagonal arrays of spots with radial tails. At even higher concentrations of substrate, larger aggregates form that seem to have a life of their own. They move slowly like slugs, with the larger slugs consuming the smaller ones. For a more complete description of such pattern formation, see Budrene and Berg (1995).

The spreading phenomena described thus far are exhibited by cells of normal size swimming through pores of soft agar, responding to chemical gradients that they generate by consumption or excretion. If cells are grown on agar with pore sizes slightly too small for the cells to penetrate (e.g., 0.5%) on a very rich medium, something very different happens. The cells get longer, produce many more flagella, and excrete a lubricant, called slime. They move rapidly outward across the surface of the agar, in parallel arrays in rafts or packs, through coordinated flagellar movement. They appear to "swarm," like bees (see Harshey, 1994). Near the edge of the swarm, groups of cells rapidly move in swirls, this way and then that, often backing up. At the very edge, they tend to line up, pointing outward. Streams of such cells colonize the entire plate within a few hours. A circularly symmetric swarm is shown in Fig. 3.8a. One in the shape of a four-leaf clover is shown in Fig. 3.8b. The cells used in Fig. 3.8a form chemotactic rings in the presence of aspartate (as in Fig. 3.2, right), but those used in Fig. 3.8b do not. This cloverleaf pattern is reproducible, but the mechanism for its formation is not known. The signals that bring about this swarm transformation are poorly understood. However, cells need to be on a surface, to grow rapidly, to excrete slime, and to be able to swim. They do not need to be chemotactic. Swarming is better known in other flagellated species, especially in *Proteus mirabilis*, where long swarm cells revert to short vegetative cells, which later develop more swarm cells, generating colonies that are terraced (Rauprich et al., 1996). Unfortunately, the word "swarm" is used in two ways: for the general phenomenon of cells swimming

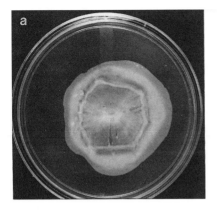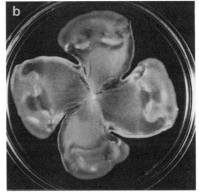

FIGURE 3.8. (a) Swarm of an *E. coli* strain deleted for genes that encode receptors for serine, ribose/galactose, and dipeptides but expressing the gene for the aspartate receptor, *tar*. (b) As in (a), but for a similar strain expressing a gene for an aspartate receptor unable to bind aspartate, *tar(T154I)*. Cells were inoculated on a Petri plate containing 0.45% Eiken agar (from Japan) and a rich growth medium (peptone, meat extract) and incubated for 16 hours at 30°C. (Burkart et al., 1998, Figs. 4A, B, reprinted with permission.)

through soft agar (as in the formation of Adler's chemotactic rings), and to denote the particular form of surface translocation just described.

References

Adler, J. 1966. Chemotaxis in bacteria. *Science* 153:708–716.

Adler, J. 1969. Chemoreceptors in bacteria. *Science* 166:1588–1597.

Adler, J. 1973. A method for measuring chemotaxis and use of the method to determine optimum conditions for chemotaxis by *Escherichia coli. J. Gen. Microbiol.* 74:77–91.

Adler, J. 1976. The sensing of chemicals by bacteria. *Sci. Am.* 234 (4): 40–47.

Adler, J. 1987. How motile bacteria are attracted and repelled by chemicals: an approach to neurobiology. *Biol. Chem. Hoppe-Seyler* 368: 163–173.

Adler, J., and B. Templeton. 1967. The effect of environmental conditions on the motility of *Escherichia coli. J. Gen. Microbiol.* 46:175–184.

Berg, H. C. 1992. Response of *Escherichia coli* to novel gradients. In: Sensory Transduction, Proc. 45th Symp. Soc. Gen. Physiol. Rockefeller University Press, New York, pp. 220–223.

Blakemore, R. 1975. Magnetotactic bacteria. *Science* 190:377–379.

Budrene, E. O., and H. C. Berg. 1991. Complex patterns formed by motile cells of *Escherichia coli*. *Nature* 349:630–633.

Budrene, E. O., and H. C. Berg. 1995. Dynamics of formation of symmetrical patterns by chemotactic bacteria. *Nature* 376:49–53.

Burkart, M., A. Toguchi, and R. M. Harshey. 1998. The chemotaxis system, but not chemotaxis, is essential for swarming motility in *Escherichia coli*. *Proc. Natl. Acad. Sci. USA* 95:2568–2573.

Frankel, R. B. 1984. Magnetic guidance of organisms. *Annu. Rev. Biophys. Bioeng.* 13:85–103.

Frankel, R. B., and R. P. Blakemore. 1991. Iron Biomaterials. Plenum, New York.

Harshey, R. M. 1994. Bees aren't the only ones: swarming in Gram-negative bacteria. *Mol. Microbiol.* 13:389–394.

Jung, K-H., Spudich, E. M., Trivedi, V. D., and Spudich, J. L. 2001. An archael photosignal-transducing module mediates phototaxis in *Escherichia coli*. *J. Bacteriol.* 183:6365–6371.

Maeda, K., Y. Imae, J.-I. Shioi, and F. Oosawa. 1976. Effect of temperature on motility and chemotaxis of *Escherichia coli*. *J. Bacteriol.* 127:1039–1046.

Nishiyama, S., I. N. Maruyama, M. Homma, and I. Kawagishi. 1999. Inversion of thermosensing property of the bacterial receptor Tar by mutations in the second transmembrane region. *J. Mol. Biol.* 286:1275–1284.

Rauprich, O., M. Matsushita, C. J. Weijer, F. Siegert, S. E. Esipov, and J. A. Shapiro. 1996. Periodic phenomena in *Proteus mirabilis* swarm colony development. *J. Bacteriol.* 178:6525–6538.

Spudich, J. L. 1998. Variations on a molecular switch: transport and sensory signalling by archaeal rhodopsins. *Mol. Microbiol.* 28:1051–1058.

Spudich, J. L., C.-H. Yang, K.-H. Jung, and E. N. Spudich. 2000. Retinylidene proteins: structures and functions from archaea to humans. *Annu. Rev. Cell Dev. Biol.* 16:365–392.

Taylor, B. L., and D. E. Koshland, Jr. 1975. Intrinsic and extrinsic light responses of *Salmonella typhimurium* and *Escherichia coli*. *J. Bacteriol.* 123:557–569.

Taylor, B. L., J. B. Miller, H. M. Warrick, and D. E. Koshland, Jr. 1979. Electron acceptor taxis and blue light effect on bacterial chemotaxis. *J. Bacteriol.* 140:567–573.

4
Individual Cells

Tracking Bacteria

If one looks through a microscope at a suspension of cells of motile *E. coli*, one is dazzled by the activity. Nearly every organism moves at speeds of order 10 body lengths per second. A cell swims steadily in one direction for a second or so (in a direction roughly parallel to its body axis), moves erratically for a small fraction of a second, and then swims steadily again in a different direction. Some cells wobble from side to side or tumble end over end. A few just seem to fidget. Given enough oxygen, the cells do this forever, even as they grow and divide. Near the middle of such a preparation, cells rapidly appear and disappear as they move in and out of focus, while at the bottom or the top they tend to spiral along the glass surface, clockwise (CW) at the bottom, counterclockwise (CCW) at the top. The speed at which the cells swim depends on how they have been grown (two to three times faster when grown on a rich medium than on a simple one), on the ambient temperature (twice as fast at body temperature than at room temperature), and on how they have been handled. Flagella are fragile and break if suspensions are subjected to viscous shear, particularly when cell densities are high (as in a centrifuge pellet). If one tries to resuspend such a pellet by flicking the centrifuge tube with one's finger, cell motility is noticeably degraded.

My interest in quantifying this motion was sparked in 1968 by a conversation with Max Delbrück, who bemoaned the fact that he did not know how to "tame" bacteria. By "tame," I finally realized, he meant monitoring the behavior of individual cells. This was what he was doing in his work on growth of the spore-bearing stalk of the fungus *Phycomyces*, simply by using a telescope. So I built a microscope that could follow the motion of individual cells of *E. coli* in three dimensions (Fig. 4.1). In essence, this is a

FIGURE 4.1. The tracking microscope, circa 1974. The lenses, mirrors, and fiber-optic assembly used to dissect the image of a cell was built into the rectangular box extending back from the top of the binocular. Just below the objective is a thermostatted enclosure containing a small chamber in which the bacteria were suspended, mounted on a platform driven by three sets of electromagnetic coils (similar to loudspeaker coils) built into the assembly at the left. (From Berg, 1978, Fig. 2).

three-dimensional direct current (DC) servo system in which errors in the position of the image of a bacterium sensed at the top of the microscope (where one normally places a camera) are used to control the position of a small chamber holding a cell suspension, so that the image (and hence the bacterium) remains fixed in the laboratory reference frame. To follow the movement of the bacterium, all one has to do is write down the position (the x, y, and z coordinates) of the chamber. It's rather like following the progress of a worm in a bucket of soil by moving the bucket so that the worm remains fixed in the reference frame of one's garden. The accelerations are so slight that neither the bacterium nor the worm knows that it is being manipulated. This is a nonperturbative measurement.

Tracking is fun. When viewed through the microscope, the cell being followed changes its orientation or its mode of vibration but remains in focus at a fixed point. The other cells drift this way and that, in apparent synchrony. One of my favorite tracks is shown in Fig. 4.2: three stereo views of the same data set, representing about 30 seconds in the life of a wild-type (behaviorally competent) cell, swimming in the absence of any chemical gradients. *E. coli* just wanders around, trying new directions at random. The smooth segments of this random walk are called "runs," and the erratic intervals are called "tumbles." During runs, the cell moves along a reasonably smooth track. During tumbles, it moves erratically in place. After a tumble, it sets off again along another smooth track, but in a new direction chosen nearly at random. Computer analysis of such data showed that run intervals are distributed exponentially, with short intervals the more probable. The lengths of successive intervals are not correlated. This is just what one finds for intervals between clicks of a Geiger counter, where emissions from a radioisotope occur with a constant probability per unit time. Not only are short intervals the more probable, they appear to be bunched. What one often calls a tumble when viewing cells by eye actually is a sequence of short runs and tumbles (which is why, in the original work, I used the word "twiddle" rather than "tumble"). The mean run interval is about 1 second, varying somewhat from cell to cell. Tumble intervals also are distributed exponentially, with a mean of about 0.1 second, but this value is the same from cell to cell.

Figure 4.3 shows the swimming speed of the cell of Fig. 4.2. The bars indicate tumbles logged by the computer. It takes the cell a while to get up to speed following a tumble, but the terminal speeds are nearly identical. The reasons for this are discussed in the next chapter.

If cells were to choose new directions at random, the distribution of turn angles would follow a sine curve, with a mean of 90 degrees. In dilute aqueous media, there is a slight preference for the forward direction, and the mean is 68 degrees. But it only takes a cell a few tumbles to forget where it has been. It does not know where it is going.

x

y

z

FIGURE 4.2. Three stereo plots of a track of one cell of *E. coli* strain AW405 (wild type for chemotaxis) viewed along the x, y, and z axes (top, middle, and bottom, respectively). To see a given plot in three dimensions, look at the left image with your left eye and the right image with your right eye, and relax your eye muscles so that the two images overlap. A stereoscope (a pair of lenses) helps. The cell was tracked in Adler's motility medium at 32°C for 29.5 seconds, and the x, y, and z outputs were digitized 12.6 times per second. The largest span across the track (e.g., from top to bottom in the middle plot) is 106 μm. There were 26 runs and tumbles; the longest run was 3.6 seconds. The mean speed was 21.2 μm/sec. (Data from Berg and Brown, 1972, Fig. 1.)

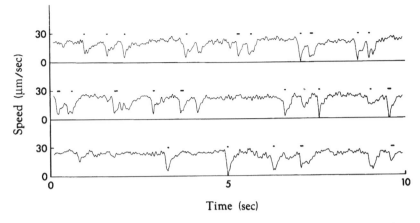

FIGURE 4.3. The speed of the cell whose track is shown in the previous figure. Tumbles occurred during the intervals shown by the bars. A strip-chart record of the output of an electronic speedometer was divided into three parts, which were stacked on top of one another. (From Berg and Brown, 1972, Fig. 2.)

Response to Spatial Gradients

How, then, do cells respond to gradients? To answer this question, we inserted one of Adler's capillary tubes (Fig. 3.5) through the side wall of a tracking chamber and followed cells in gradients of serine and aspartate. Given Engelmann's demonstration of the shock reaction, we had expected that *E. coli* would shorten runs that are unfavorable. The result proved to be exactly the opposite. *E. coli* extends runs that are favorable (that carry cells up the gradient of an attractant) but fails to shorten runs that are not (that carry cells down such a gradient). The random walk of Fig. 4.2 becomes biased, and the bias is positive. The bias is large enough to enable a cell to move up a gradient at about 10% of its run speed. There is no correlation between the change in direction generated by a tumble and the cell's prior course; tumbles have precisely the same effect whether a cell swims in a gradient or not, they just occur with different frequencies. Thus, if life gets better, *E. coli* swims farther on the current leg of its track and enjoys it more. If life gets worse, it just relaxes back to its normal mode of behavior. *E. coli* is an optimist.

Response to Temporal Gradients

The next question was whether cells respond to spatial or tempo-ral stimuli. That is, is a favorable run extended because the cell finds more attractant near its nose than near its tail, or because the concentration goes up as it moves along? Recall that the answer for *Chromatium* was temporal. When Engelmann passed his hand between the light source and the microscope stage, all the cells in the field of view backed up; when he exposed cells in a hanging drop to carbon dioxide, they backed up regardless of their orientation relative to the surface of the drop. We decided to answer this question for *E. coli* by a method that did not expose cells to spatial inhomogeneities, such as those encountered during mixing of chemicals or diffusion into the surface of a drop. We found an enzyme, available commercially, that would convert an innocuous substance into a chemical attractant. The reaction was reversible, so alternatively the attractant could be destroyed. Thus, no matter where a cell might be or where it might be headed, it would always find the concentration of the attractant rising or falling. When the attractant was generated, all the runs got longer. When it was destroyed, the cells failed to respond. The response to the positive temporal gradient was large enough to account for the results obtained in spatial gradients (Brown and Berg, 1974).

The question of whether cells respond to spatial or temporal stimuli had been considered earlier in a simpler way by Macnab and Koshland (1972), who rapidly mixed suspensions of cells and attractants and recorded the response under a microscope using stroboscopic illumination. Cells suddenly exposed to a positive step of serine (0 to 0.8 mM) swam smoothly (without tumbling) for up to 5 minutes. Cells exposed to a negative step (1 to 0.24 mM) tumbled incessantly for about 12 seconds. These exper-iments showed that *E. coli* (actually *Salmonella*) senses temporal stimuli. Technically, this was true not because the cells responded, but because the responses to positive and negative steps were dif-ferent (of opposite sign), even though the spatial homogeneities to which the cells were exposed during mixing were roughly the same. *E. coli* does not encounter temporal stimuli of this magni-tude when swimming in spatial gradients in nature. Unless there is a strong source (e.g., a fine capillary tube) and a strong sink (e.g., a large surrounding pond), spatial gradients are rapidly smoothed out by diffusion. In any event, cells do not swim fast enough to generate large temporal stimuli. Such stimuli saturate the response: in the mixing experiments, cells either swam without tumbling or tumbled incessantly, although much longer in the former than in the latter case. What one measures is the time required for the cells to recover (i.e., to return to a mode in which they run and tumble). However, such stimuli have proved quite useful for probing the chemotaxis machinery.

References

Berg, H. C. 1978. The tracking microscope. *Adv. Opt. Elect. Microsc.* 7:1–15.

Berg, H. C., and D. A. Brown. 1972. Chemotaxis in *Escherichia coli* analysed by three-dimensional tracking. *Nature* 239:500–504.

Brown, D. A., and H. C. Berg. 1974. Temporal stimulation of chemotaxis in *Escherichia coli*. *Proc. Natl. Acad. Sci. USA* 71:1388–1392.

Macnab, R. M., and D. E. Koshland, Jr. 1972. The gradient-sensing mechanism in bacterial chemotaxis. *Proc. Natl. Acad. Sci. USA* 69: 2509–2512.

5
Flagellar Motion

Rotation

Whether a cell runs or tumbles depends on the direction of rotation of its flagella, but the story turns out to be rather complicated. A tumble involves not only a change in the direction of rotation of one or more of the flagellar filaments, but also a sequence of changes in their handedness and pitch.

During a run, after a cell has gotten up to speed, all of the filaments rotate in the same direction, usually counterclockwise (as seen by an observer behind the cell). Each filament turns individually, but they go around side by side. If you think that the filaments should tangle up or tie in knots, take two thin rods—aluminum welding rods work fine—and wrap them into identical shallow helices by bending them around a pipe. They need not be precisely helical, they just need to be the same shape. Then hold the helices at one end, side by side, and roll them between your fingers. If the helices are left-handed (spiral to the left as they extend away from you) and you turn them clockwise (counterclockwise as seen by someone looking at them from the other end), they will turn smoothly, in parallel. They will do so even if they cross over one another, because the points of crossover travel away from you and are shed at the distal end (Macnab, 1977). If you turn the helices the other way and the wires happen to cross over, then the bundle will jam when the point of crossover reaches your hand. It takes considerable force to break such jams, and the wires rattle as they snap over one another. So the motion in one direction is smooth and quiet and in the other direction rough and noisy.

One of the initial arguments for flagellar rotation (Berg and Anderson, 1973) was the fact that a small amount of bivalent antifilament antibody would jam flagellar bundles, while a large amount of monovalent antifilament antibody (a bivalent antibody cut in two) had no effect. Evidently, the bivalent antibody worked by cross-linking one filament to another, preventing the rotation, while the monovalent antibody simply made each filament thicker. Another argument was that two cells linked together by their flagella (actually their hooks) counterrotated. A filament is joined to the drive shaft of the motor at its base by a short flexible coupling called the proximal hook. Mutants had been found in which these hooks were abnormally long and filaments were largely absent. Such cells were nonmotile. However, when antihook antibody was added, these cells formed pairs that counterrotated.

This assay was perfected by Silverman and Simon (1974), who cemented filaments (or hooks) to glass using antifilament (or antihook) antibody. If only one filament (or hook) was tethered in this way, the cell body spun at speeds of about 10 Hz (revolutions per second), alternately clockwise or counterclockwise (CW or CCW), changing directions about once per second. Such a tethered cell is shown in Fig. 5.1, spinning CCW.

The correspondence between CCW rotation and runs, on the one hand, and CW rotation and tumbles, on the other, was then established by tethering cells and adding attractants or repellents (Larsen et al., 1974). When a large amount of chemical attractant was added, the cells spun exclusively CCW for several minutes, just as swimming cells ran exclusively in the mixing experiments of Macnab and Koshland (1972). If, instead, a large amount of repellent was added, the cells spun exclusively CW, but only

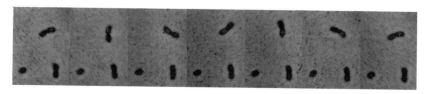

FIGURE 5.1. Three cells of *E. coli* wild-type strain AW405 tethered to a glass coverslip by a single flagellar filament (top) or simply stuck to the glass (bottom) shown at intervals of 0.1 second beginning at the left. The tethered cell (a long cell about to divide) completes one revolution counterclockwise (CCW). Its axis of rotation is near the right end of the image on the left. Note that the concave side of the cell leads and the convex side lags: the cell is rotating (like a pinwheel) not gyrating (like your arm when you wave it in a circle). (From Berg, 1976, Fig. 1.)

for several seconds, again just as cells tumbled in the mixing experiments.

Filament Shape

Flagellar filaments are relatively stiff, but they can switch between distinct polymorphic forms. Four of these forms are shown in Fig. 5.2. The normal filament is left-handed, and the semicoiled and curly filaments are right-handed. The filament is a polymer of a single protein called flagellin, whose molecules can bond in two different ways. They appear as 11 rows of protofilaments along the surface of a cylinder, as shown in Fig. 5.3. When the flagellin molecules are bonded in one way, the row is short; when they are bonded in the other way, the row is long. If all of the protofilaments in a flagellar filament are identical, the filament is straight. There are two kinds of straight filaments, short or long, but the difference in their lengths is relatively small. However, if some of the protofilaments are short and others are long, the filament is

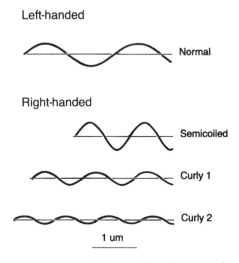

FIGURE 5.2. Drawing of four different flagellar waveforms, each with a contour length of $4\,\mu$m. A filament of this length contains about 8000 molecules of flagellin (Hasegawa et al., 1998). The normal filament is left-handed, and the semicoiled, curly 1, and curly 2 filaments are right-handed. The normal and curly 1 filaments have the same overall length. Bar, $1\,\mu$m. (Adapted from Calladine, 1978.)

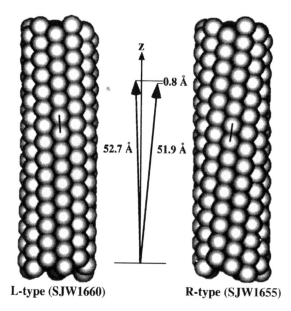

L-type (SJW1660)   **R-type (SJW1655)**

FIGURE 5.3. The surface lattice of L- and R-type straight flagellar filaments. The spacing between flagellin subunits along an 11-start helix (a protofilament) of the R-type is 0.08 nm smaller than between corresponding subunits of the L-type. L and R refer to the handedness of the filament twist. The SJW numbers are numbers of bacterial strains. The distances are measured at a radius of 4.5 nm and are shown magnified in the middle of the drawing. (Namba and Vonderviszt, 1997, Fig. 19, reprinted with permission.)

helical: the short protofilaments run along the inside of the helix. The different shapes shown in Fig. 5.2 arise from different numbers of adjacent protofilaments of a given type. Transformations between these polymorphic forms can be driven by changes in protein structure (i.e., by mutations in the flagellin gene), by changes in the composition of the surrounding medium [e.g., in pH (acidity) or ionic strength (salt content)], or by mechanical twist (i.e., by torsion).

Tumbling

Until recently, it was thought that tumbles occur when all of the flagellar motors switch from CCW to CW, even though experiments in which motors were studied in isolation (in the absence of large stimuli and without interacting filaments) suggested that each motor switches independently. The resolution to this puzzle was found on labeling flagella with a bleach-resistant fluorescent dye and recording their motion in a fluorescence microscope using strobed laser illumination (Turner et al., 2000).

The simplest case is a cell with a single flagellar filament (Fig. 5.4). A transformation from normal to semicoiled is seen in fields 4 to 10, from semicoiled to curly 1 in fields 12 to 18, and from curly 1 back to normal in fields 24 to 30. The cell swam into the field of view moving toward the 7 o'clock position and left the field of view moving toward the 5 o'clock position. Most of this change in direction occurred while the filament was partially in the semi-

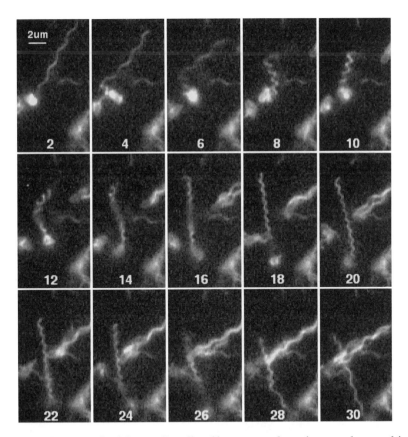

FIGURE 5.4. A cell with one flagellar filament undergoing a polymorphic transformation. The video recording was made at 60 Hz, but only every other field is shown. The numbers are in units of 1/60 second. Note the scale bar (2 μm). (From Turner et al., 2000, Fig. 6.)

coiled form (fields 4 to 12). Evidently, the flagellar motor switched from CCW to CW after field 2 and back again after field 22.

This pattern also occurs in cells with several filaments, where the tumble is generated by changes in the direction of rotation of as few as one or as many as all of the filaments. Generally, the more filaments that are involved, the larger the change in direction. As the filaments regain their normal conformation, they rejoin the normal bundle. Figure 5.5 shows a cell with two flagellar filaments, only one undergoing polymorphic transformations. Note the curly 1 filament wrapping around the normal filament and rejoining the bundle as it reverts to the normal conformation, fields 17 to 20. This cell swam into the field of view moving toward

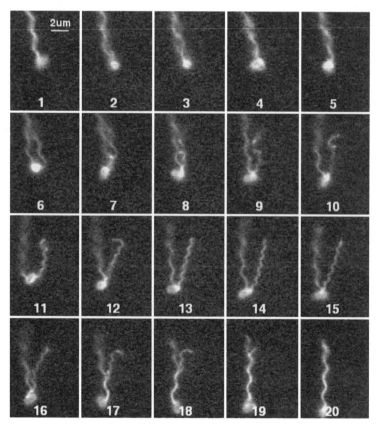

FIGURE 5.5. A cell with two flagellar filaments, only one undergoing polymorphic transformations. (From Turner et al., 2000, Fig. 7.)

the 5 o'clock position and left the field of view moving toward the 6 o'clock position.

The sequence of normal, semicoiled, curly 1, and then back to normal is observed most frequently, as summarized in Fig. 5.6. The change in direction of the cell body generally occurs early on, while the filament is partially in the semicoiled form. This explains why the time required for the cell to change direction, indicated in Fig. 4.3 by the horizontal bars, is substantially shorter than the time required for the cell to get back up to speed, indicated by the corresponding speed trace. The cell in Fig. 5.6 starts out along its new path being pushed by a curly 1 filament spinning CW and a

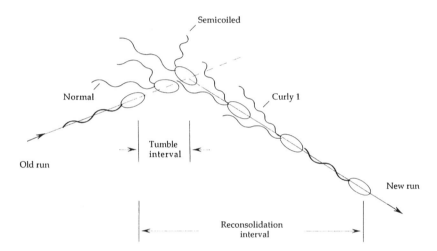

FIGURE 5.6. A schematic drawing of the events that usually occur during a tumble. A cell with a bundle of two flagellar filaments is shown swimming from left to right. The cell alters course as the motor driving one filament changes its direction of rotation and the filament undergoes a normal to semicoiled transformation. This change in course defines the tumble interval, which, according to both the tracking and video data, takes 0.14 second, on average. As the cell begins to move along its new track, the filament undergoes a semicoiled to curly 1 transformation. Both the normal and curly 1 filaments generate forward thrust, but the curly one at a smaller magnitude. Finally, after the direction of flagellar rotation changes again, the filament reverts to normal. As it does so, it rejoins the bundle, and the cell resumes its initial speed. The time from the initial disruption of the bundle to its reconsolidation is defined as the reconsolidation interval. According to the video data, this takes 0.43 second, on average.

normal filament spinning CCW. This propulsion is not as efficient as when both filaments are normal and spinning CCW.

High-speed video recording reveals that transformations from normal to semicoiled or curly 1 are triggered by changes in direction of flagellar rotation from CCW to CW, as expected, while transformations back to normal are triggered by changes in direction from CW back to CCW. But it also is possible for filaments of different kinds to spin backward without changing their overall shape.

In earlier work with swimming cells studied by dark-field microscopy, Macnab and Ornston (1977) observed the curly 1 transformation. Hotani (1982), working with isolated filaments fixed to glass at one end, was able to generate both semicoiled and curly 1 transformations, by flow of a viscous medium. In dark field, an enormous amount of light is scattered by the cell body, so Hotani had an easier task than Macnab and Ornston. The problem of scattering from the cell body is eliminated by the fluorescence technique.

Complications notwithstanding, we are left with the remarkable conclusion that the behavior of the cell depends on the direction of rotation of rotary motors that drive propellers that change their handedness and pitch.

References

Berg, H. C. 1976. Does the flagellar rotary motor step? In: Cell Motility. R. Goldman, T. Pollard, T. Rosenbaum, editors. Cold Spring Harbor Laboratories, Cold Spring Harbor, NY, pp. 47–56.

Berg, H. C., and R. A. Anderson. 1973. Bacteria swim by rotating their flagellar filaments. *Nature* 245:380–382.

Calladine, C. R. 1978. Change in waveform in bacterial flagella: the role of mechanics at the molecular level. *J. Mol. Biol.* 118:457–479.

Hasegawa, K., I. Yamashita, and K. Namba. 1998. Quasi- and nonequivalence in the structure of bacterial flagellar filament. *Biophys. J.* 74: 569–575.

Hotani, H. 1982. Micro-video study of moving bacterial flagellar filaments III. Cyclic transformation induced by mechanical force. *J. Mol. Biol.* 156:791–806.

Larsen, S. H., R. W. Reader, E. N. Kort, W. Tso, and J. Adler. 1974. Change in direction of flagellar rotation is the basis of the chemotactic response in *Escherichia coli. Nature* 249:74–77.

Macnab, R. M. 1977. Bacterial flagella rotating in bundles: a study in helical geometry. *Proc. Natl. Acad. Sci. USA* 74:221–225.

Macnab, R. M., and D. E. Koshland, Jr. 1972. The gradient-sensing mechanism in bacterial chemotaxis. *Proc. Natl. Acad. Sci. USA* 69:2509–2512.

Macnab, R.M., and M.K. Ornston. 1977. Normal-to-curly flagellar transitions and their role in bacterial tumbling. Stabilization of an alternative quaternary structure by mechanical force. *J. Mol. Biol.* 112:1–30.

Namba, K., and F. Vonderviszt. 1997. Molecular architecture of bacterial flagellum. *Q. Rev. Biophys.* 30:1–65.

Silverman, M., and M. Simon. 1974. Flagellar rotation and the mechanism of bacterial motility. *Nature* 249:73–74.

Turner, L., W. S. Ryu, and H. C. Berg. 2000. Real-time imaging of fluorescent flagellar filaments. *J. Bacteriol.* 182:2793–2801.

6
Physical Constraints

For a microscopic organism living in water, such as *E. coli*, constraints imposed by physics are immediate and compelling. These limit the means by which cells are able to swim, define the distance that they must move to determine whether life is getting better or worse, and set the time scale for their behavioral response. To appreciate what *E. coli* has accomplished, we need to look at some of the physics that *E. coli* knows.

The physics that looms large in the life of *E. coli* is not the physics that we encounter, because we are massive and live on land, while *E. coli* is microscopic and lives in water. To *E. coli*, water appears as a fine-grained substance of inexhaustible extent, whose component particles are in continuous riotous motion. When a cell swims, it drags some of these molecules along with it, causing the surrounding fluid to shear. Momentum transfer between adjacent layers of fluid is very efficient, and to a small organism with very little mass, the viscous drag that results is overwhelming. As a result, *E. coli* is utterly unable to coast: it knows nothing about inertia. When you put in the numbers (Berg, 1993) you find that if a cell swimming 30 diameters per second were to put in the clutch, it would coast less than a tenth of the diameter of a hydrogen atom! And a tethered cell spinning 10 Hz would continue to rotate for less than a millionth of a revolution. But cells do not actually stop, because of thermal agitation. Collisions with surrounding water molecules drive the cell body this way and that, powering brownian motion (Brown, 1828). For a swimming cell, the cumulative effect of this motion over a period of 1 second is displacement in a randomly chosen direction by about 1 μm and rotation about a randomly chosen axis by about 30 degrees. As a consequence, *E. coli* cannot swim in a straight line. After about 10 seconds, it drifts off course by more than 90 degrees, and thus forgets where it is going. This sets an upper limit on the time available for a cell to decide whether life is getting better or worse. If it cannot decide within about 10 seconds, it is too late. A lower limit is set by the time required for the cell to count enough molecules of attractant or repellent to determine their concentrations with adequate precision. The number of receptors required for this task proves surprisingly small, because the random motion of molecules to be sensed enables them to sample different points on the cell surface with great efficiency.

Viscosity

If you take a thin wire, hold it vertically, and drop it in a viscous medium, it falls straight down at some velocity, v. If, instead, you drop it horizontally, it falls straight down at about half that velocity, $v/2$. The viscous drag on the wire (the force per unit velocity that resists its motion) depends on the orientation: it is about twice as large when the wire moves sideways than when it moves lengthwise. As a consequence, if you drop the wire slantwise, say tilted downward to the right, it falls slantwise to the right. A formal analysis of a closely related problem, in which a wire is held slantwise and pulled straight downward, is shown in Fig. 6.1.

  E. coli carries out this experiment by wrapping the wire into a helix and turning it about the helical axis, as shown, for example, in Figs. 5.4 and 5.5. The helix behaves like a series of wire segments pulled downward or upward, slantwise, in such a way that the forces generated by each segment in a direction parallel to the helical axis add up, providing the thrust that moves the cell body forward. If the cell (with its flagella) swims at a constant speed (does not accelerate or decelerate), it does not experience any net force; therefore, the thrust generated by the rotating helix must be balanced by the drag on the cell body. The same argument applies to rotation: the torque exerted by the flagellar motors on the filaments must be balanced by counterrotation of the cell body. However, since the body is relatively large, it turns relatively slowly. So when *E. coli* swims, the flagellar bundle spins one way on the order of 100 Hz, while the cell body rolls the other way on the order of 10 Hz; the cell with its flagella moves forward at speeds of order 10 body lengths per second. For a human being, 10 body lengths per second is about 40 miles per hour!

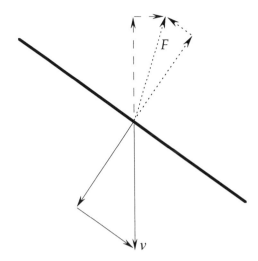

Figure 6.1. A thin wire held slantwise and pulled downward through a viscous medium at velocity v. This velocity can be decomposed into components perpendicular to the wire and parallel to the wire, as shown below the wire. The drag due to the perpendicular component is twice as great per unit velocity as the drag due to the parallel component, as shown by the dotted lines above the wire. The net drag is F. It is not vertical but is tilted to the right, so it has a horizontal as well as a vertical component, as shown by the dashed lines. The horizontal component tends to move the wire to the right. If the wire were a segment of a rotating helix, this component would provide thrust. The vertical component opposes v, and thus determines the power required to move the filament. If the wire were a segment of a rotating helix, this component would contribute to the torque required to rotate the helix. For the orientation shown (55 degrees from vertical), the ratio of the horizontal to the vertical components (0.354) is maximum.

Reynolds Number

In a viscous medium, the ratio of the forces required to accelerate masses (inertial forces) to the forces required to generate shear (viscous forces) is called the Reynolds number, R. For a swimming creature, $R = lv\rho/\eta$, where l is the size of the creature, v is its velocity, ρ is the density of the medium, and η is the viscosity of the medium (a coefficient that characterizes its resistance to shear). For *E. coli* swimming full speed in water, $R \approx 10^{-5}$ (1/100,000). For a human paddling slowly in a swimming pool, $R \approx 10^5$ (100,000). We are much bigger (l is much bigger) and we move much more rapidly (v is much bigger). So, in a certain sense, our experience in water differs from that of *E. coli* by a factor of 10^{10}. Our inertia is large, and it is easy for us to push off and coast from one side of the pool to the other. If you want to model what life is like for *E. coli* on a larger scale (by scaling up l and/or v), then you also must scale up η (work with a highly viscous medium). So use glycerol or corn syrup or a thick silicone oil, and don't move things too rapidly. This restriction was not clearly understood until the work of Ludwig (1930), whose contribution was forgotten by the time the problem was taken up again by Taylor (1952).

Ludwig noted a remarkable thing about motion at a low Reynolds number. If a pattern of displacements is reversed in time (neglecting diffusion), all elements of the system return to their initial positions, cell and fluid alike. The rate at which these displacements are carried out does not matter. Ludwig illustrated this point by imagining a creature with two rigid oars attached to the cell body by hinges, as shown in Fig. 6.2. The organism strokes its oars rapidly downward and returns them slowly upward. At a low Reynolds number, the cell body moves rapidly upward and then slowly downward, returning to its initial position. At a high Reynolds number, on the other hand, it moves farther during the power stroke than during the recovery stroke. There are microscopic unicellular algae that look somewhat like this cell (e.g., *Chlamydomonas*). However, they move their flagella in different ways during the power and recovery strokes: far from the cell body during the power stroke and close to the cell body during the recovery stroke (as in the human breast stroke). This motion is cyclic but not reciprocal; that is, the pattern is not reversed in time. Therefore (as Ludwig noted), it works at a low Reynolds number. The flagellar motion exhibited by *E. coli* also is cyclic: as long as the flagellar filaments turn steadily counterclockwise, the cell swims steadily forward.

Vivid images of this world were evoked by Purcell (1977) in an article titled, "Life at low Reynolds number." Suppose, for example, that you are immersed in a swimming pool full of molasses and are allowed to move parts of your body no faster than the hands of a clock? According to Purcell, "If under those ground rules you are able to move a few meters in a couple of weeks, you may qualify as a low Reynolds number swimmer." This world, while rather baffling to us, is one that *E. coli* knows intimately.

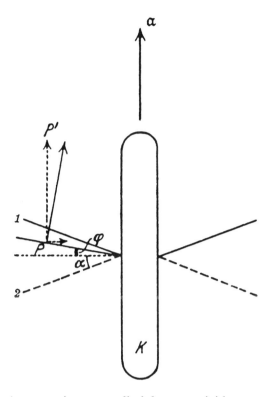

FIGURE 6.2. An organism propelled by two rigid oars, according to Ludwig (1930, Fig. 2). The oars move up and down between positions 1 and 2. A microscopic organism of this kind would just jiggle up and down. A macroscopic one, on the other hand, could swim by pulling the oars rapidly downward and returning them slowly upward. The arrows and Greek symbols in the figure relate to Ludwig's analysis of the problem, not examined here.

Diffusion

It is more difficult to model the utterly random motion due to thermal agitation. Whereas one can study the motion of macroscopic objects at low Reynolds numbers by working in highly viscous media, it is difficult to scale up a diffusion coefficient. There are no liquids with viscosities much lower than that of water, and work in gases is not practical because of perturbations due to gravity, notably, sedimentation and convection. It is easier to use a microscope and think small. The major take-home lesson is this: diffusive transport over small distances is very efficient, while diffusive transport over large distances is very inefficient. Diffusion times increase as the square of the distance. Thus, a small molecule in water can diffuse the width of *E. coli* (1 μm) in a few milliseconds. To diffuse the width of your finger (1.5 cm), it takes about a day.

To see how this comes about, consider a one-dimensional random walk. An ensemble of small creatures live on the x-axis and step with probability 1/2 to the right (+) or to the left (−) a distance δ every τ seconds. A record of the progress of six such creatures after 10 steps would look something like this:

Steps taken	Distance moved	Distance squared
---+--+---	−6δ	36δ^2
++-+++----	0	0
+++-++--+-	+2δ	4δ^2
-------+++	−4δ	16δ^2
-+-+-++--+	0	0
++-+-+-+-+	+2δ	4δ^2

This list was generated by flipping a coin. Some creatures drift to the right, some to the left, but on average—one needs a larger list to prove this—they go nowhere. The mean displacement for this list is $\langle x \rangle = -\delta$, where the brackets denote an ensemble average. But the creatures have spread out, and one can get a measure of this by computing their mean-square displacement (the average of the square of the displacement), which for this list is $\langle x^2 \rangle = 10\delta^2$. The mean-square displacement increases linearly with the number of steps (see Berg, 1993, Chapter 1). For example, if you break this list in half and treat it as 12 creatures each taking five steps, you will find a mean-square displacement 6.3δ^2, which is about half as large as before. Now, if t is the running time for the experiment, the number of steps is t/τ, so $\langle x^2 \rangle = (t/\tau)\delta^2 = (\delta^2/\tau)t$. The coefficient that characterizes step distances and step times is commonly written $D = \delta^2/2\tau$, which gives $\langle x^2 \rangle = 2Dt$. This is the mean-square displacement for one dimension. Similar equations can be written for motion along the y and z axes. If the motions along the x, y, and z axes are statistically independent (the usual case), then the mean-square displacement in two dimensions is $\langle x^2 + y^2 \rangle = 4Dt$, and the mean-square displacement in three dimensions is $\langle x^2 + y^2 + z^2 \rangle = 6Dt$.

D is called the diffusion coefficient. It depends on the size of the particle (and to a lesser extent, its shape), the viscosity of the medium in which the particle is immersed, and the temperature.

For a small molecule in water $D \approx 10^{-5}\,\text{cm}^2/\text{sec} = 10^{-9}\,\text{m}^2/\text{sec}$. So when I said a small molecule can diffuse the width of *E. coli* in a few milliseconds, what I really meant was $t = \langle x^2 \rangle / 2D \approx (10^{-6}\,\text{m})^2/(2 \times 10^{-9}\,\text{m}^2/\text{sec}) = 5 \times 10^{-4}\,\text{sec}$. That is, if a molecule starts out at one side of the cell at time 0, the chances are pretty good that it will reach the other side within a millisecond. But the chances are equally good that it will have gone a similar distance in the opposite direction (neglecting the impediment of the cell wall). *The diffusion coefficient characterizes a spreading distance, not a velocity. Indeed, there is no such thing as a diffusion velocity: because of the square, it takes a set of diffusing particles four times as long to spread twice as far.* To diffuse 1.5 cm, $t = (1.5 \times 10^{-2}\,\text{m})^2/(2 \times 10^{-9}\,\text{m}^2/\text{sec}) = 1.1 \times 10^5\,\text{sec} = 1.3$ days. For globular-shaped particles in water, D is proportional to $T/a\eta$, where T is the absolute temperature, a is the radius of the particle, and η is the viscosity of water (which is smaller at higher temperatures).

A simulation of a two-dimensional random walk is shown in Fig. 6.3. Diffusive transport over small distances is very efficient: the plotter pen tended to explore some regions of space rather thoroughly, returning to the same point many times before wandering away for good. Diffusive transport over large distances is very inefficient: when the plotter pen did wander away, it did so blindly, with no inkling of where it had been or where it might go. As a result, some parts of the plot are filled in, and others are quite empty.

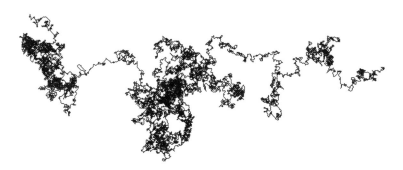

FIGURE 6.3. An x,y plot of a two-dimensional random walk of 21,537 steps. At each step a computer flipped a coin twice and moved the plotting pen diagonally, to the right upward for +,+; to the right downward for +,−; to the left upward for −,+; and to the left downward for −,−. The first 18,050 steps of this walk are shown in Berg (1993, Fig. 1.4).

As we have seen, when *E. coli* swims, it picks directions at random. Therefore, it also diffuses. The step lengths for a motile cell are much longer than those due to thermal agitation, but they do not occur as often. The translational diffusion coefficient for a wild-type cell is much larger than that for a nonmotile cell, roughly $D = 4 \times 10^{-10}$ m²/sec, as compared to 2×10^{-13} m²/sec. But even a smooth-swimming mutant executes a random walk, because rotational diffusion carries the cell off course. The same kind of coin-flipping experiment with increments in angle yields a mean-square angular displacement about one axis $\langle \theta^2 \rangle = 2D_r t$, where D_r is a rotational diffusion coefficient. For globular-shaped particles in water, D_r is proportional to $T/a^3 \eta$. As noted earlier, this mechanism carries *E. coli* off course by about 90 degrees in 10 seconds. As a result, the translational diffusion coefficient for the smooth-swimming mutant works out to about $D = 2 \times 10^{-9}$ m²/sec, roughly 5 times that of the wild-type cell. To learn more, see Berg (1993, Chapters 4, 6).

Diffusion of Attractants or Repellents

Diffusion of attractants or repellents sets a lower limit on the distance (and thus the time) that a cell must swim to outrun diffusion (to reach greener pastures), as well as on the precision with which the cell, in a given time, can determine concentrations. Diffusion of attractants or repellents also determines the number of receptors of a given kind that the cell needs to carry out these measurements. If a cell remains in one place for time t, it will sample molecules that come from a distance of order $(Dt)^{1/2}$, where D is their diffusion coefficient. If the cell swims at velocity v during time t, it will be displaced a distance of order vt. If it is to go far enough to find out whether life is getting better or worse, it must outrun diffusion. This implies $vt > (Dt)^{1/2}$, or $t > D/v^2$. For *E. coli* swimming 30 μm/sec, $t > (10^{-9} \text{m}^2/\text{sec})/(3 \times 10^{-5} \text{m/sec})^2 \approx$ 1 sec. This time is approximately equal to the mean run length. Recall that when a cell responds to gradients of attractants or repellents, it tends to extend runs rather than shorten them. Presumably, it does this because it can learn more by doing so. Short runs are not very informative.

If attractants or repellents are absorbed by a moving cell, there are fewer available at the back than at the front, but the difference proves to be small (Berg and Purcell, 1977). Nevertheless, this difference is large enough to rule out a mechanism in which a rapidly moving cell compares counts in the front with those in the back, that is, in which it makes spatial comparisons. The apparent gradient generated by the motion is several hundred times steeper than gradients encountered during chemotaxis. As a result, were the cell to choose a new direction at random, any direction would be deemed favorable. In other respects, however, the spatial mechanism is viable: a stationary cell could obtain the precision required to detect small differences in concentrations at its poles, simply by counting molecules for a relatively long time. The moving cell does so by comparing counts as a function of time, that is, by making temporal comparisons.

It is possible to estimate the time required for a cell to measure the concentraton of molecules with a given precision. Assume that the cell can count molecules in its own volume, a^3, where a is its linear dimension (10^{-6} m). The result of one such count is $a^3 C$, where C is the mean concentration of molecules in its environment. Sampling of this kind is governed by the Poisson distribution, and the standard deviation is equal to the square-root of the mean (Berg, 1993, p. 90). Therefore, the uncertainty in the count is $(a^3 C)^{1/2}$, yielding a precision (the standard deviation divided by the mean) of $(a^3 C)^{-1/2}$. For $E.\ coli$ in, say, $1\,\mu$M aspartate, $(a^3 C)^{-1/2}$ $\approx [(10^{-6}\text{m})^3\ (6 \times 10^{20}\ \text{molecules/m}^3)]^{-1/2} = 0.04$, or 4%. The cell can do better by waiting for the molecules that it has counted to diffuse away and for another set to diffuse in. If this happens, the two counts will be statistically independent. The required waiting time is of order a^2/D, where D is the diffusion coefficient. If the cell continues this process for time t, the total count will increase by a factor $t/(a^2/D) = Dt/a^2$, yielding a final count $DaCt$, with precision $(DaCt)^{-1/2}$. For $t = 1$ sec, $a = 10^{-6}$ m, and $D = 10^{-9}$ m²/sec, $Dt/a^2 = 10^3$, yielding a precision of about 0.1%.

To determine whether the concentration is going up or down, the cell has to make two such measurements and take the difference. It will not be able to make an informed decision unless this difference is larger than its standard deviation. Since things improve as $t^{1/2}$, it would appear that the cell might work to arbitrarily high precision, simply by counting for very long times. But as we have seen, rotational brownian movement of the cell body sets an upper limit of order $t = 10$ sec. To correct its course, the cell must deal with the recent past, not the distant past. So, for the counts to be large enough, C cannot be too small. For a cell swimming $30\,\mu$m/sec integrating counts over periods of 1 sec, a precision of 0.1% (as estimated for $1\,\mu$M aspartate, above) is sufficient for sensing a gradient with a decay length of about 2 cm. For a more rigorous discussion of the counting problem, see Berg and Purcell (1977).

There is an additional wrinkle. The cell can only count molecules if they bind to a receptor. The chemotaxis machinery inside the cell monitors the occupancy of these receptors. A molecule of attractant diffuses around until it finds an empty binding site, sticks for a short time, and then diffuses away. The ratio of the off- to the on-rates is known as the dissociation constant, K_d, which equals the concentration, in moles per liter, at which the receptor occupancy is one-half. This is the concentration at which the receptors are most sensitive to fractional changes in concentration. To work at concentrations large enough for adequate precision, the receptors for the best attractants (e.g., aspartate or serine) have dissociation constants in the micromolar range. If the on-rates are diffusion limited, the dwell times (inverse off-rates) turn out to be about 10^{-4} sec. Therefore, some device within the cell must compute the fraction of time that a receptor is occupied. Molecules continuously bind to the receptor and diffuse away, sticking for a time quite short compared to the time required for the cell to complete a single measurement.

How many receptors of a given kind must a cell have to count a substantial fraction of the molecules that impinge on its surface? As evident from the preceding discussion and Fig. 6.3, it takes a given molecule a relatively long time to reach a specific region of space. But once it is there, it explores that region rather thoroughly. Once a molecule encounters the cell surface, it tends to collide with that surface hundreds or thousands of times before it wanders away for good. As a result, it has an excellent chance of encountering a specific binding site. One can show that *E. coli* can do about half as well with a few thousand receptors of a given kind as it would do were its entire surface dedicated to that one specific task (see Berg, 1993, pp. 30–33). As a result, the cell has room for many different kinds of receptors (or transporters), each working at reasonable efficiency. This is a boon, not a constraint. Without benefits of this kind, microscopic life would not be possible.

Recapitulation

Since *E. coli* is more familiar with this world that we are, let me repeat. Flagellar filaments are long, thin, and helical, because motion is dominated by viscous rather than inertial forces: thrust is generated by viscous drag. A cell is unable to swim in a straight line, because rotational perturbations due to brownian movement knock it off its path. Long runs are more effective for exploring the environment than short ones, because they allow the cell to outrun diffusion of the molecules that it needs to count. Rapidly moving cells must sense chemical gradients temporally rather than spatially, because comparisons between concentrations in front or behind are overwhelmed by diffusive currents due to their motion. Finally, the precision with which a cell can make temporal comparisons is limited by statistical fluctuations. The counting statistics improve with the square root of the product of the concentration and the integration time. A chemical cannot be sensed at an arbitraily low concentration because the integration time required would be prohibitively long.

References

Berg, H. C. 1993. Random Walks in Biology. Princeton University Press, Princeton.

Berg, H. C., and E. M. Purcell. 1977. Physics of chemoreception. *Biophys. J.* 20:193–219.

Brown, R. 1828. A Brief Account of Microscopical Observations on the Particles Contained in the Pollen of Plants; and on the General Existence of Active Molecules in Organic and Inorganic Bodies. Richard Taylor, London.

Ludwig, W. 1930. Zur Theorie der Flimmerbewegung (Dynamik, Nutzeffekt, Energiebalanz). *Z. Vgl. Physiol.* 13:397–504.

Purcell, E. M. 1977. Life at low Reynolds number. *Am. J. Phys.* 45:3–11.

Taylor, G. I. 1952. The action of waving cylindrical tails in propelling microscopic organisms. *Proc. R. Soc. Lond. A* 211:225–239.

7
Optimal Control

To find out whether *E. coli* really knows what it is doing, Ed Purcell and I thought hard about the theory of chemoreception—I was the straight man—and concluded that its cells can sense temporal gradients about as well as any other device of similar size could possibly do (Berg and Purcell, 1977). And then my students and I looked more closely at how changes in tumble probability actually depend on the concentrations of attractants or repellents.

Time Resolution

To do this, we needed to stimulate cells in a known way and record responses on a time scale smaller than 1 second. This is hard to do by adding chemicals and mixing. Also, the problem is complicated by the fact that the response is stochastic: the probability of tumbling changes, but intervals between tumbles remain exponentially distributed. So one needs lots of data.

In recent work (e.g., Jasuja et al., 1999), ultraviolet light is used to cleave a photosensitive molecule. One of the fragments released is a chemical attractant (e.g., the amino acid aspartate). This allows one to generate concentration jumps on the millisecond time scale. We chose, instead, to use iontophoretic pipettes, developed earlier by others to stimulate receptors at the neuromuscular junction. This allows one to generate pulses as well as jumps, but on a somewhat longer time scale. The limit is the time required for a small molecule to diffuse from the tip of the pipette to a cell a few micrometers away, about 20 msec. Our target was either a tethered cell, fixed to glass by a single flagellar filament, or a filamentous cell linked via a single flagellum to an inert marker, as shown in Fig. 7.1. The first setup was used to probe the chemotactic response at the level of a single flagellar motor (Block et al., 1982) and the second to learn how signals are transmitted

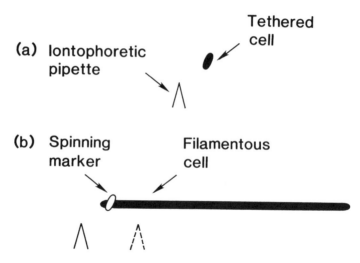

FIGURE 7.1. Stimulation with iontophoretic pipettes. (a) The tip of a pipette near a cell tethered to glass, such as the cell in Fig. 5.1. (b) The tip of a pipette either off one end or along the body of a filamentous cell linked by a single flagellum to an inert marker. This linkage was made via an abnormally long hook, called a polyhook, to polyhooks of a cell of normal size that had been treated with a chemical fixative (glutaraldehyde). Filamentous cells were obtained by growing normal cells in the presence of an antibiotic similar to penicillin (cephalexin) or by using mutants defective in septation. Such cells have a single cytoplasmic compartment.

intracellularly from the receptors to the flagella (Segall et al., 1985). The pipette was filled with a solution similar to the medium in which the cells were suspended containing, in addition, an attractant [e.g., aspartate (Fig. 3.1) or its nonmetabolizable analog α-methylaspartate]. At neutral pH, either amino acid has a net charge of -1, so it is expelled from the pipette when the electrical potential difference between the inside and the outside of the pipette is negative.

Impulse Responses

One can learn a great deal about a mechanical system by exciting it with a brief pulse. If, for example, you kick a sign post, it will wobble back and forth at a frequency that depends on its stiffness and mass and relax back to its initial quiescent state with a time constant that depends on the rate at which mechanical energy is dissipated. You will get essentially the same result whether you wear a boot or a tennis shoe. If the system is linear, that is, if the way it responds to a new stimulus does not depend on how it is responding to past stimuli, the response to the impulse allows one to predict the response to any stimulus. Decompose the stimulus of interest into a sequence of impulsive stimuli of different magnitudes, weight the corresponding impulse responses by these magnitudes, and add them up.

The same is true for biochemical systems. If you kick the aspartate receptor by loading it up with ligand for a fraction of a second, the reactions set in motion by that change will play themselves out until the cell returns to its initial quiescent state. In practice, this takes about 4 seconds (Fig. 7.2). The impulse response for *E. coli* is biphasic. The probability that the motor spins counterclockwise rises from the baseline soon after the onset of the pulse, reaches

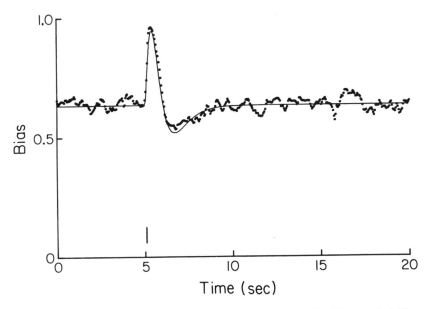

FIGURE 7.2. Impulse response of wild-type *E. coli* cells. The probability that a cell spins counterclockwise (the bias) is plotted as a function of time; the smooth curve is a fit to a sum of exponentials. Pulses of aspartate or α-methylaspartate were applied beginning at 5.06 seconds (vertical bar). The graph was constructed from 378 trials comprising 7566 flagellar reversals obtained with 17 cells. (From Segall et al., 1986, Fig. 1).

a maximum about 0.4 second later, crosses the baseline 1 second after the pulse, reaches a minimum at 1.5 second, and finally returns to the baseline at about 4 seconds. The areas of the positive and negative lobes of the response are equal (Segall et al., 1986).

From this analysis, it follows that wild-type cells exposed to stimuli in the physiological range (stimuli that do not saturate the response) make short-term temporal comparisons extending 4 seconds into the past. Stimuli received during the past second are given a positive weighting, and stimuli received during the 3 seconds before that are given a negative weighting, and the cells respond to the difference. The cells count molecules over a substantial time span—this improves the precision of the count—and then ask (within the time limit set by rotational brownian movement) whether the concentration is going up or down. This provides an optimum solution to the measurement problem, a solution that is matched to the constraints imposed by the physics discussed in Chapter 6. Simpler strategies, for example, one in which a cell sets its tumbling probability on the basis of measurements of the local concentration, do not work (Schnitzer et al., 1990).

The impulse response for a negative pulse (one that lowers the concentration of an attractant or raises the concentration of a repellent) is similar to the response shown in Fig. 7.2, except that it is of opposite sign (Block et al., 1982). Experiments with cells exposed to ramps of either sign indicate that thresholds for positive stimuli are small, while those for negative stimuli are large (Block et al., 1983). However, once these thresholds are crossed, equal increments in ramp rate generate equal increments in rotational bias, until the ramps are so steep that saturation occurs. Thus, if a cell has fully adapted, small negative stimuli are ignored. Evidently, this is why cells fail to respond when swimming down spatial gradients of attractants or when exposed to attractants destroyed enzymatically (see Chapter 4).

If one looks at these data in the frequency domain, one finds that the sensory system behaves as a bandpass filter, with its response maximally tuned to frequencies of a few tenths of a Hz, approximately equal to those encountered when cells move up and down in a spatial gradient, as shown in Fig. 7.3. Thus, E. coli has matched its sensory system to the signals that it needs to analyze.

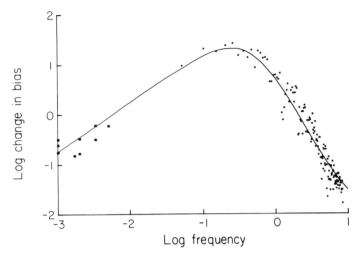

FIGURE 7.3. Impulse and ramp data viewed in the frequency domain. The change in bias resulting from variations in the concentration of an attractant (α-methylasparate) at different frequencies are plotted in a log-log scale. Data at the left were obtained from exponentiated sine-wave stimuli, while data at the right were obtained from the Fourier transform of the impulse response. The smooth line is a similar transform of the smooth curve of Fig. 7.2. (From Segall et al., 1986, Fig. 3B.)

Simulations of the Biased Random Walk

Given the impulse response, it is possible to simulate the biased random walk. Pick a run velocity and let the cell move by displacing it, say, every 0.01 second. Weight the recent and more distant past using the impulse response in the manner described earlier. If the output of this computation is negative, that is, if the concentration has been decreasing for some time, ignore the result and determine whether the cell should tumble by picking at random from an exponential distribution with a mean of 1 second. If the output of this computation is positive, determine whether the cell should tumble by picking at random from an exponential distribution with a larger mean (one with an exponent decreased in linear proportion to that output). If the cell is tumbling, determine whether it should run by picking from an exponential distribution with a mean of 0.1 second. In either case, if a new run is called for, pick the change in angle from the old to the new run at random from a distribution peaked in the forward direction (Berg and Brown, 1972, Fig. 3). Finally, add the effects of brownian rotation by giving the cell a small kick in angle every iteration. When watching such tracks evolve on a computer screen, one gets the impression of a bloodhound following a scent. The cell sniffs about (with the bias close to the baseline), picks up the spoor, and then howls up the gradient. Eventually, rotational brownian motion carries it off the track, and it is forced to sniff about again. Most of the progress up the gradient appears to occur in long runs.

Intracellular Signaling

Experiments of the sort sketched in Fig. 7.1b were used to study the range of the intracellular signal that couples the receptors to the flagella (Segall et al., 1985). Stimuli delivered at one end of a filamentous cell did not affect the response at the other end. There was no evidence for long-range signaling, as would be expected, for example, were the receptors to signal the flagella by changing membrane potential. Motors near the pipette responded, whereas those far away did not. The response of a given motor decreased with distance, but it did so less sharply when the pipette was moved along the cell surface (to the right in Fig. 7.1b) than when it was moved out into the external medium (to the left in Fig. 7.1b). This implies that there is an internal signal, but that its range is short (only a few micrometers). The data could not be fit by models in which the receptor simply releases or binds a small molecule or in which a receptor-attractant complex diffuses to the flagellar motor. However, they could be fit by a model in which the signal is a ligand or a small protein that is activated by the receptor and inactivated as it diffuses through the cytoplasm. As we will see in Chapter 9, this molecule proved to be a small protein, CheY, which is active when phosphorylated and inactive when not.

References

Berg, H. C., and D. A. Brown. 1972. Chemotaxis in *Escherichia coli* analysed by three-dimensional tracking. *Nature* 239:500–504.

Berg, H. C., and E. M. Purcell. 1977. Physics of chemoreception. *Biophys. J.* 20:193–219.

Block, S. M., J. E. Segall, and H. C. Berg. 1982. Impulse responses in bacterial chemotaxis. *Cell* 31:215–226.

Block, S. M., J. E. Segall, and H. C. Berg. 1983. Adaptation kinetics in bacterial chemotaxis. *J. Bacteriol.* 154:312–323.

Jasuja, R., Y. Lin, D. R. Trentham, and S. Khan. 1999. Response tuning in bacterial chemotaxis. *Proc. Natl. Acad. Sci. USA* 96:11346–11351.

Schnitzer, M., S. Block, H. C. Berg, and E. Purcell. 1990. Strategies for chemotaxis. *Symp. Soc. Gen. Microbiol.* 46:15–34.

Segall, J. E., S. M. Block, and H. C. Berg. 1986. Temporal comparisons in bacterial chemotaxis. *Proc. Natl. Acad. Sci. USA* 83:8987–8991.

Segall, J. E., A. Ishihara, and H. C. Berg. 1985. Chemotactic signaling in filamentous cells of *Escherichia coli. J. Bacteriol.* 161:51–59.

8
Cellular Hardware

Before we look at the biochemical machinery that enables cells to count molecules, to compare counts made at different times, and to use these results to control the direction of flagellar rotation, we need to know more about cell architecture.

Body Plan

E. coli is a single-celled organism with a multilayered wall (Fig. 8.1). First, there is a thin outer membrane made of lipopolysaccharide, with the sugar chains pointing outward, penetrated by holes due to proteins, called porins. This membrane blocks the passage of most lipid-soluble molecules, but it allows the passage of water-soluble things up to about twice the size of sucrose. Next, there is a porous gauze-like layer of peptidoglycan that gives the cell its rigidity and cylindrical shape. This structure resists the turgor pressure generated when the cell finds itself in a medium of low osmotic strength. The peptidoglycan is immersed in an aqueous layer, called the periplasm, containing a variety of smaller molecules, including a number of proteins that either bind molecules that interest the cell or destroy molecules that pose a threat. Finally, there is a cytoplasmic (phospholipid) membrane similar to the membranes that enclose human cells, spanned by proteins involved in generation of energy, transport of materials, and sensory transduction. This is the main permeability barrier that enables the cell to retain chemicals—DNA, RNA, protein, and a variety of water-soluble molecules of lower molecular weight— essential for life. The multilayered cell wall is on the order of $0.03\,\mu\text{m}$ thick. Unlike most human cells, which contain a number of membranous organelles, including a nucleus, and a variety of rope-like and tubular cytoskeletal structures, the cytoplasm of *E. coli* is a quasi-homogeneous soup.

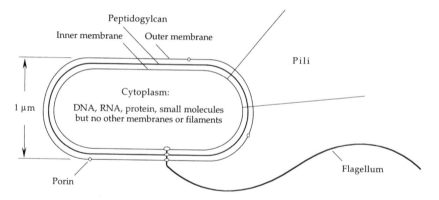

FIGURE 8.1. A schematic diagram of *E. coli*. One flagellum, two type 1 pili, and three porin molecules are shown. A typical cells has four flagella, either zero or a hundred or so pili, and thousands of porins. The flagellum comprises a long helical filament, a short proximal hook, and a basal body. The basal body is embedded in the cell wall. The spacing between the inner and outer membranes is shown larger than scale by a factor of about 4.

Phospholipids are molecules with polar (water-loving) head groups containing phosphate and long hydrocarbon (oily) tails. They form bilayer membranes, with the oily tails on the inside and the head groups on the outside. The oily layer blocks the passage of water-soluble molecules. In bacteria, there also are a variety of polymers in which phospholipids are combined with sugars to form lipopolysaccharide, and peptides are combined with sugars to form peptidoglycan (also called murein). In both cases the sugars appear in long chains. Some of the most effective antibiotics (e.g., penicillin) interfere with the synthesis of these components (e.g., peptidoglycan). Since human cells do not have such walls, we are not harmed by these antibiotics (unless we happen to be allergic). To learn more, see Seltmann and Holst (2002).

The most remarkable molecule in the cytoplasm of *E. coli* is DNA, a double-helical chain about 1.4 mm long (nearly 1000 times longer than the cell) in the form of a closed loop. When cells grow rapidly, there is more than one copy per cell, because the molecule is being replicated at more than one place. The DNA has about 4.6×10^6 base pairs. Since 3 base pairs specify one amino acid, it can code for 1.5×10^6 amino acids (ignoring regions of DNA that specify RNA or are required to bind proteins that turn genes on and off). A typical polypeptide has a molecular weight of about 50,000, comprising 400 amino acids. So the DNA can code for about 4000 polypeptides. The coding region for each polypeptide is called a gene. The DNA of *E. coli* K-12 has been sequenced and found to code for 4288 polypeptides (Blattner et al., 1997). Surprisingly, the pathogenic strain O157:H7 has been found to code for 32% more (Perna et al., 2001). The functions of only about 60% of the original set of gene products are known. Fewer than 2% are involved in bacterial chemotaxis.

When a polypeptide is made, the relevant DNA sequence is copied as a messenger RNA (mRNA, often short-lived), which is read by a large RNA-protein particle called a ribosome. This links specific amino acids end to end. These are supplied by transfer RNAs (tRNA) that recognize successive 3-base codons in the mRNA. In an electron micrograph of a sectioned *E. coli*, the cytoplasm appears granular, because there are many ribosomes, each about $0.03\,\mu m$ in diameter. Regions in which the DNA is more concentrated appear less granular, because ribosomes tend to be excluded. Dissolved in a finer matrix are mRNA, tRNA, a variety of proteins, and chemicals of low molecular weight.

Why Cells?

It is worth pausing to consider why all free-living things, including *E. coli*, are cells or are made up of cells. A cell is a relatively small isolated device that can import foodstuffs and export wastes, grow, and replicate. The isolation is essential both for chemistry and for genetics.

For reactions to occur, chemicals need to find one another. The rate of a reaction such as $A + B \rightarrow C$ is proportional to the product of the concentrations of A and B. If A and B are both diluted by a factor of 1000, the reaction rate goes down by a factor of one million. So one needs a concentrated medium in which to do biochemistry. Bacteria are the earliest cells that we know anything about, and they are relatively small. The time required for a small molecule to diffuse across a cell $1\,\mu m$ in diameter is a matter of milliseconds (see Chapter 6). So early cells did not need any specialized machinery for moving goods from one place to another. Thermal agitation would do.

Specific chemical reactions are catalyzed by enzymes. In the earliest forms of life, these probably were made from ribonucleic acid (RNA). Now they are made from proteins, large molecules that also serve as structural elements (see below). Every time a cell divides, the DNA that encodes these proteins replicates. Occasionally, mistakes are made, and a different structure is specified. If the change is beneficial, so that the new cell is more likely to survive, then the mistake can propagate. But this can happen only if the molecule that carries the genetic information is packaged together with the product that it specifies. So evolution works because DNA is able to reap the rewards imposed by natural selection. In an earlier world, before cells were invented, the RNA that catalyzed essential reactions must have been self-replicating.

More on Proteins

By weight, *E. coli* is about 70% water, 1% inorganic ions, and the balance organic molecules, most of high molecular weight. Proteins are polymers made up of precise linear sequences of 20 different kinds of amino acids—amino acids have a molecular weight averaging about 120, an amino group (as in ammonia), distinctive side chains, and a carboxylic-acid group (as in vinegar) that can form peptide bonds linking one subunit to another, as we saw in Fig. 3.1. Proteins contain one or more long polypeptide chains. For example, hemoglobin, a protein well known to us but not to *E. coli*—hemoglobin carries oxygen in our blood—has four polypeptide chains. A space-filling model is shown in Fig. 8.2.

Polypeptide chains tend to wind up in a helix, called an α-helix, or line up side by side (parallel or antiparallel) in a sheet, called a β-pleated sheet. Thus, polypeptides also can be represented by ribbon diagrams, in which the shape of the ribbon indicates the local conformation of the chain, as shown in Fig. 8.3.

Growth

When *E. coli* grows, it gets longer and then divides in the middle. In a sense, it is immortal, because the mother is replaced by two daughters, essentially identical to the daughters of the previous generation. *E. coli* is haploid; it has only one chromosome. The fate of each cell is determined by a single DNA double helix. Except for mutations, which occur spontaneously for a given gene at the rate of about 10^{-6} per generation, all the molecules of DNA in a given set of descendents are identical. If fed well and held at

FIGURE 8.2. Hemoglobin, the protein that carries oxygen in our blood, and the sugar glucose, shown on the same scale. This is a space-filling illustration; bonds between individual atoms are not shown. Hemoglobin is a compact globular structure of diameter 5.5 nm and molecular weight about 65,000 in which four polypeptide chains are packed together in a tetrahedral array. (From Goodsell, 1993, reprinted with permission.)

the temperature of the human gut (37°C), *E. coli* divides every 20 minutes. In a medium with only one carbon source (e.g., glucose), it takes longer, about 2 hours. The extra time is required for the cell to synthesize all of the other organic molecules that it needs. A generation time of 20 minutes is prodigious. If we start with one cell at noon today, there will be $2^3 = 8$ cells at 1:00 o'clock, and 2^{72} $= 4.7 \times 10^{21}$ cells at noon tomorrow. Since each cell has a volume of about $10^{-18} m^3$, the volume of cells at noon tomorrow will be $4.7 \times 10^3 m^3$, i.e., a cube about 17 meters = 55 feet on a side! In practice, this does not happen, because the cells are not provided with enough food. However, it explains why, when, say, 100 cells are dispersed on the surface of hard nutrient agar, one soon obtains 100 mounds of cells (colonies) each a millimeter or so in diameter, or why, on soft agar, the progeny of a single cell soon populate the entire plate. It is this speed of replication that makes

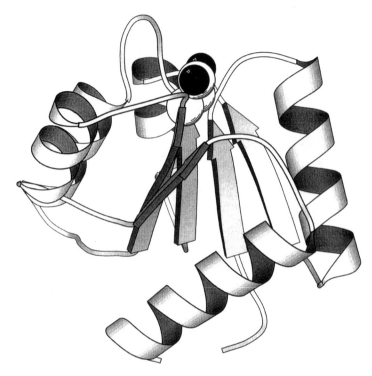

FIGURE 8.3. A ribbon diagram of CheY, a small protein that couples receptors to the flagella (see Chapter 9). CheY becomes active when phosphorylated. The phosphorylation site, aspartate-57, is shown at the top in space-filling format, with the side-chain oxygens in black. A central five-stranded parallel β-sheet is sandwiched between five α-helices. (Falke et al., 1997, Fig. 14A, reprinted with permission.)

bacterial genetics such a pleasure. Start your experiment today and get an answer tomorrow.

External Organelles

E. coli has at least three kinds of extracellular organelles. These include two kinds of fibers, called pili, that extend out from the cytoplasmic membrane. The first, very thin and straight, called type 1 pili (or fimbriae), enable the bacteria to adhere to a variety of surfaces, including cells of the intestinal epithelium. There can be hundreds per cell. A second, called the sex pilus, can bind to other cells and retract, drawing the cells together. This enables cells carrying F plasmids (small autonomously replicating circular pieces of DNA) to transfer part of their DNA to other cells. Cells with F plasmids make one such pilus. Some cells used for studies of bacterial behavior have type 1 pili and others do not; few have sex pili.

The extracellular organelle of greatest interest here is the bacterial flagellum. This organelle has three parts, a basal body (a reversible rotary motor) embedded in the cell wall (beginning within the cytoplasm and ending at the outer membrane), a short proximal hook (a flexible coupling or universal joint), and a long helical filament (a propeller). The filament is about $0.02\,\mu m$ in diameter. Normally it is shaped as a left-handed helix with a wavelength (pitch) of about $2.3\,\mu m$ and a diameter of $0.4\,\mu m$. If a cell is not subjected to viscous shear—filaments are easily broken—its filaments can be up to $10\,\mu m$ long. The number of flagella vary from cell to cell. The average number is about four.

To learn more about basic molecular and cellular biology, see Alberts et al. (2003). For an earlier more succinct view, see Kendrew (1966).

References

Alberts, B., D. Bray, K. Hopkin, A. Johnson, J. Lewis, M. Raff, K. Roberts, and P. Walter. 2003. Essential Cell Biology. An Introduction to the Molecular Biology of the Cell, Second Edition. Garland, New York.

Blattner, F. R., G. Plunkett, C. A. Bloch, et al. 1997. The complete genome sequence of *Escherichia coli* K-12. *Science* 277:1453–1462.

Falke, J. J., R. B. Bass, S. L. Butler, S. A. Chervitz, and M. A. Danielson. 1997. The two-component signaling pathway of bacterial chemotaxis: a molecular view of signal transduction by receptors, kinases, and adaptation enzymes. *Annu. Rev. Cell Dev. Biol.* 13:457–512.

Goodsell, D. S. 1993. The Machinery of Life. Springer-Verlag, New York.

Kendrew, J. C. 1966. The Thread of Life. G. Bell & Sons, London.

Perna, N. T., G. Plunkett III, V. Burland, et al. 2001. Genome sequence of enterohaemorrhagic *Escherichia coli* O157:H7. *Nature* 409:529–533.

Seltmann, G., and O. Holst. 2002. The Bacterial Cell Wall. Springer, Berlin.

9
Behavioral Hardware

Components

Suppose you discovered a computerized factory turning out small cars, and you wanted to know how those cars were assembled and how they functioned. One way to identify essential components would be to remove those components one at a time and then characterize the resulting defects. For example, if you removed the drive shaft, the engine would run but the drive wheels would not turn, so the car would be paralyzed. If you knew the computer program, you could do this at will by removing the instructions for fabrication or assembly of drive shafts. If you did not know those instructions, or indeed even what a car might be, you could still learn a great deal by changing the program at random (e.g., by making mutants). This is how things proceeded in the early days of bacterial chemotaxis. One mutagenized cells, isolated mutants with interesting defects (e.g., cells with flagella that failed to spin), and then mapped the gene. Given the gene, one could identify the gene product. Now things are much easier. The genetic program is known in detail, and one can modify it in any way that one desires. For example, one can amplify a specific gene by using the polymerase chain reaction (PCR), change its sequence at will, and put it back into the chromosome by homologous recombination. Or one can paste the gene into a multicopy plasmid behind a strong promoter and express the gene product at high concentrations. The techniques for doing these things are straightforward, but outside the scope of this book. The essential point is that one can use genetics to identify and manipulate components (proteins) involved in any cellular process, including bacterial chemotaxis. The parts required for motility and chemotaxis are described in this chapter. The way in which the genetic map is read and these gene products are assembled is described in the next chapter.

   Mutations affecting chemotaxis have specific phenotypes (behavioral defects), and genes tend to be named for those defects. In some cases, where the gene was identified first in another context (e.g., the gene for the maltose binding protein, *malE*, involved in maltose transport), the name is foreign to chemotaxis. In most cases, however, the abbreviation is closer to home; for example, *trg*, for taxis toward ribose or galactose; or *cheA*, for the first gene identified with a generally nonchemotactic phenotype—*che* cells swim but do not make chemotactic rings or respond in the capillary assay; or *motB*, for the second gene identified with a defective motility phenotype—*mot* cells make flagella, but these flagella fail to spin; or *fliF*, for a gene required for flagellar synthesis. The early flagellar mutants were named *flaA*, *flaB*, etc., but the alphabet proved too short, so now they are called *flg*, *flh*, *fli*, and *flj*, depending on their location on the chromosome (Iino et al., 1988). When one refers to the gene product, that is, the protein specified rather than the gene, the first letter is capitalized and italics are not used. Names appear in this form in the parts lists given in the appendix. Table A.1 in the appendix lists components involved in chemoreception, Table A.2 lists components involved in signal processing, and Table A.3 lists components involved in motor output. Components of different types or subtypes are listed alphabetically.

Signaling Pathway

The sensory transduction pathway is shown schematically in Fig. 9.1, where the information flow is from left to right. The same system is depicted four times: each row of the figure illuminates a different aspect of the mechanism, as explained in the figure legend. The basic scheme, shown in row 1, is typical of a number of so-called two-component signaling pathways in bacteria, in which information, embodied by a phosphate group, is passed from a histidine kinase to an aspartate kinase (Parkinson and Kofoid, 1992). These components are named for the amino acid residues that carry the phosphate. The histidine kinase is coupled to a sensor, and the aspartate kinase (also called a response regulator) is coupled to an effector. In pathways involving gene regulation, the effector interacts with a particular transcriptional control element. In chemotaxis, there are two effectors, the rotary motor and a methylesterase, an enzyme that demethylates the receptor, as shown in rows 2 and 3. The response regulator that interacts with the motor diffuses to its base, where, if phosphorylated, it binds and increases the probability of clockwise (CW) rotation. The response regulator that activates the methylesterase comprises the N-terminal domain of the same protein; if phosphorylated, it activates the C-terminal domain, which carries the catalytic site. During adaptation to rising concentrations of attractants, the receptor is methylated by a methyltransferase; during adaptation to falling concentrations of attractants, the receptor is demethylated by the methylesterase. The different proteins that make up this system are named in row 4, where the example given is for taxis toward the sugar maltose and the amino acid aspartate.

It is worth noting the location of this hardware within the cell plan shown in Fig. 8.1. Tar spans the inner membrane. Aspartate or MalE binds transiently at its periplasmic end. Aspartate finds Tar and maltose finds MalE by diffusing through the porins in the outer membrane. MalE is confined to the periplasmic space. CheR binds transiently to the C-terminal end of Tar, within the cytoplasm at a site located in between the inner membrane and the innermost end of the receptor. CheW and CheA bind at the innermost end of the receptor to form a stable complex. Che B and CheY bind to CheA until phosphorylated, and then they diffuse freely within the cytoplasm. FliM is a component at the cytoplasmic face of the flagellar motor. There are several motors distributed at random along the sides of the cell, each of which penetrates the cell wall.

So, we have two kinds of sophisticated protein machines, both embedded in the inner membrane: the receptor complex and the flagellar motor. They are coupled by diffusion of a small cytoplasmic protein, activated by phosphorylation.

Receptor Complex

The receptor named in row 4 of Fig. 9.1, Tar (for taxis toward aspartate or away from certain repellents), is in a class of receptors known as methyl-accepting chemotaxis proteins (MCPs), all of which span the cytoplasmic membrane (see Table A.1). Another class of membrane receptors (not shown in the figure) phosphorylate their substrates and transport the derivatives. They are part of the phosphotransferase system (PTS). A novel receptor, Aer, related to the MCPs, carries a flavin adenine dinucleotide

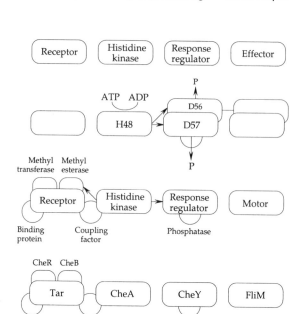

FIGURE 9.1. The sensory transduction pathway, shown in block form, repeated four times. Information flows from left to right.

Row 1: *Basic scheme.* An attractant molecule (the ligand) binds to a receptor at the outer surface of the inner membrane (in the periplasmic space). This changes the level of activity of a cytoplasmic histidine kinase that phosphorylates two response regulators (also called aspartate kinases). These, in turn, act on two effectors. The effector for the first response regulator is the flagellar motor. The effector for the second response regulator is an enzyme (a methylesterase) that targets receptor methyl groups. Interactions between the response regulators and their effectors change the probabililty that the motor spins clockwise and the activity of the methylesterase, respectively.

Row 2: *Phosphate flow.* The histidine kinase catalyzes the transfer of inorganic phosphate from adenosine triphosphate (ATP) to its own histidine-48 (H48), leaving ATP as the diphosphate (ADP). The first response regulator (shown in front) catalyzes the transfer of phosphate from H48 to its own aspartate-57 (D57), and the second response regulator (shown in the back) catalyzes the transfer of phosphate from H48 to its own aspartate-56 (D56). Hydrolysis of D57-P (removal of the phosphate) is accelerated by another enzyme (a phosphatase). Hydrolysis of D56-P occurs spontaneously; it is not catalyzed by a phosphatase. The effector for the second response regulator (the methylesterase) is the C-terminal domain of the same protein, so it is shown connected to the response regulator by a horizontal line.

that serves as a redox sensor; however, this receptor is not methylated (Taylor et al., 1999).

As noted above, Tar, CheW, and CheA form a complex, a supramolecular machine, shown schematically in Fig. 9.2. Early studies of isolated components suggested that each complex comprises two molecules of Tar, two molecules of CheW, and two molecules of CheA (or possibly one molecule of $CheA_S$ and one of $CheA_L$; see Table A.2, note c); however, the exact stoichiometry is still a matter of debate. CheA and CheW bind at the extreme intracellular end of the Tar dimer, and CheR binds to a pentapeptide at the Tar C-terminus. CheB binds to a domain in CheA downstream of H48, as does CheY.

Tar is made up of a string of α-helical segments, denoted $\alpha 1$ through $\alpha 9$ (Kim et al., 1999). Helix $\alpha 1$ (also called TM1, for transmembrane 1) starts at the inner face of the cytoplasmic membrane, crosses this membrane, and extends into the periplasm, where with helices $\alpha 2$ to $\alpha 4$ it forms an antiparallel 4-helix bundle. Helix $\alpha 4$ (also called TM2) goes back through the membrane and is connected by a linker region that includes $\alpha 5$, to the remaining helices, $\alpha 6$ to $\alpha 9$. These fold back onto one another and with helices $\alpha 6$ to $\alpha 9$ of the other copy of Tar form a second antiparallel 4-helix bundle. Helices $\alpha 6$ and $\alpha 9$ and their

----

Row 3: *Additional components.* The response regulator/methylesterase has been redrawn as a single component at the left, in contact with the receptor, with which it interacts. The arrows from the histidine kinase indicate phosphate transfer from H48 to D57 and D56, as before. Additional components include periplasmic binding proteins, required for chemotaxis toward certain sugars or dipeptides and away from nickel, a coupling factor required for activation of the histidine kinase, and a methyltransferase that methylates the receptor. The phosphatase, shown earlier, is now labeled as such.

Row 4: *Complete system.* This is shown for chemotaxis toward the disaccharide maltose and the amino acid aspartate. The receptor Tar binds aspartate and the maltose binding protein when the latter carries maltose. It also binds the nickel binding protein when it carries Ni^{2+} (not shown). CheR is the methyltransferase, CheB the methylesterase (both domains), CheW the coupling factor, CheA the histidine kinase, CheY the response regulator that when phosphorylated binds to the flagellar motor, CheZ the phosphatase that accelerates the dephosphorylation of CheY-P, and FliM the component at the base of the flagellar motor to which CheY-P binds.

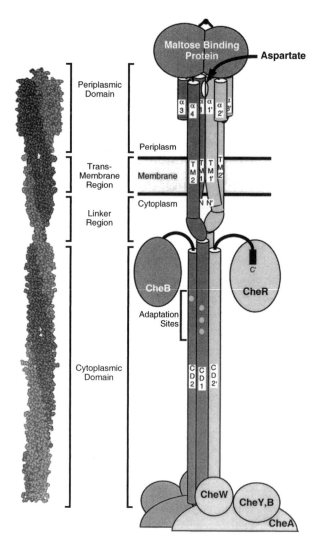

FIGURE 9.2. *Left*: A space-filling model of the Tar receptor dimer. It is 38 nm long by 2.5 nm in diameter (at the intracellular end). *Right*: A cartoon of the Tar receptor complex, including CheB, CheR, CheA, and CheW, drawn to the same scale. The dimeric association is stable; one set of components is more darkly shaded. Aspartate and the maltose binding protein are in rapid association-dissociation equilibrium with Tar. Aspartate binds in a cleft between α-helices $\alpha 1$ and $\alpha 1'$. The maltose binding protein has two domains connected by a hinge. The hinge closes when maltose binds, and then the protein binds at the periplasmic tip of the Tar dimer. Che R, CheB, and CheY also come and go, with the affinity between CheA and the phosphorylated products, CheY-P and CheB-P, substantially reduced. CheR binds to a short peptide, part of a flexible

homologs form a methylation domain—the sites of methylation, called adaptation sites in Fig. 9.2, are glutatmate side chains—while helices $\alpha7$ and $\alpha8$ and their homologs form a signaling domain. These domains are highly conserved between members of the MCP class.

Figure 9.2 shows the maltose binding protein and aspartate interacting with this receptor complex. The maltose binding protein has two domains connected by a multistrand hinge, as indicated in the figure. Maltose binds in a cleft between the two domains, and the hinge closes. Following this event, the protein binds at the extreme extracellular end of the Tar receptor complex. Aspartate binds in a cleft between $\alpha1$ and its homolog $\alpha1'$. Two binding sites are possible, but binding at one site markedly reduces the affinity of binding at the other. A great deal of work has gone into determining the change in structure that carries information about binding across the cytoplasmic membrane. The majority view is that it is a surprisingly small (0.16 nm) piston movement (toward the cytoplasm) of helix $\alpha4$ of only one of the Tar subunits (Falke and Hazelbauer, 2001). But not all agree (Kim et al., 2002).

In any event, when ligand binding occurs, the activity of CheA is reduced, and the rate of CheY-P production falls. Since CheY-P is hydrolyzed, its concentration falls, and less CheY-P binds to the base of the flagellar motor (to FliM). Therefore, the motor is more likely to spin counterclockwise (CCW), and runs are extended. In addition, the change of structure in the methylation domain increases the activity of the methyltransferase, CheR, and the reduced activity of CheA decreases the concentration of CheB-P, the active form of the methylesterase. Therefore, more glutamate side chains are methylated. This acts like a volume control to compensate for the effect of chemoattractant binding, and the activity of CheA returns to its initial value. Thus, if cells are exposed to a step-change in the concentration of maltose or

chain at the C-terminus of Tar, in a position where it can reach the methylation sites. These are shown as lighter gray dots (one set of four, labeled Adaptation Sites). CheB was thought to bind in a similar way (as shown) but is now known to bind more tightly to CheA. TM, transmembrane helix; CD, cytoplasmic domain. Proteins other than Tar are shown as ellipsoids, with CheA truncated to save space. [Courtesy of Joseph Falke, who used the space-filling model of Kim et al. (1999).]

aspartate, they eventually adapt. Addition of methyl groups is a relatively slow process, regulated by the shape of the Tar substrate. Removal of methyl groups, on the other hand, is a relatively fast process, catalyzed by CheB-P and regulated by CheA. At steady state, the rates of methylation and demethylation balance, and methylation levels are constant.

The receptor complex shown in Fig. 9.2 is a remarkable system that acts as a comparator. The output of this comparator (the kinase activity) depends on the difference between the time-average occupancies of the receptor binding sites and the level of methylation. The kinase is activated if the methylation level is relatively high and inactivated if it is relatively low. Changes in the occupancies of the receptor binding sites are very fast, and reflect the present concentrations of ligands. Changes in the levels of receptor methylation, on the other hand, are relatively slow, and reflect the past concentrations of ligands. Thus, the cell is able to make temporal comparisons. If the concentration of attractant increases steadily with time, for example, as it does when a cell swims up a spatial gradient of aspartate, the receptor occupancy rises accordingly, and the system goes out of balance. The methylation level lags behind receptor occupancy, and the kinase is slightly inactivated. Therefore, favorable runs are extended. When the cell swims down a spatial gradient of attractant, the receptor occupancy falls accordingly. But now, since demethylation is rapid, the methylation level drops rapidly, as well, and the system remains more closely in balance. Thus, the cell tends to tumble as often as it does in the absence of a stimulus. (But this is not the whole story, because, as discussed earlier, there is a threshold for repellent stimuli below which no behavioral changes can be detected.)

CheY

The structures of all of the components shown in Fig. 9.1, row 4 (except for the N-terminal domain of FliM) have been determined by x-ray diffraction or nuclear magnetic resonance. For a review of some of this work, see Falke et al. (1997). We already have seen one example taken from that source: Fig. 8.3 shows a ribbon diagram of CheY. The autocatalytic aspartate kinase pocket is at the top, formed by loops at the end of the β-sheet, with aspartate-57 shown in space-filling spheres. Overlapping domains of the surface of the molecule interact specifically with other components of the transduction system, with CheA on the left, CheZ on the right, and FliM in the middle. This is one of the smallest components of the chemotaxis system (molecular weight 14,000), a protein optimized for diffusion. Since CheY-P is unstable, its structural analysis has required major feats. The structure of activated CheY bound to the N-terminal 16 residues of its target, FliM, has been determined by x-ray diffraction of a stable beryllium fluoride derivative (Lee et al., 2001). As for Tar, the differences in structure between inactive and active forms appear to be subtle.

Flagellar Motor

The flagellar motor is shown schematically in Fig. 9.3. The electron micrograph is of the part of the motor attached to the hook that survives extraction with neutral detergents. The image has been rotationally averaged: it is what you would see if you could look through this part of the motor as it rotates. Structures outside the cell wall include the filament (the propeller), which can be up to about 10 μm long, and the hook (a flexible coupling, or universal joint). Structures embedded in the cell wall comprise the basal body and include several rings and a rod. The outer pair of rings (FlgH, called the L-ring, for lipopolysaccharide, and FlgI, called the P-ring, for peptidoglycan), is thought to serve as a bushing that gets the rod (FlgB, FlgC, FlgF, and FlgG) through the outer membrane. The rod serves as the drive shaft. Other bacteria that do not have an outer membrane, so-called gram-positive cells, do not have the outer pair of rings. And mutants of *E. coli* in which these rings are missing are motile, provided the hook protein (FlgE) is overproduced. Therefore, the L and P rings are not involved in torque generation. The inner pair of rings, formerly called M (for membranous) and S (for supramembranous) are now called MS, because they are the product of a single gene, *fliF*. An additional ring (called the C-ring, for cytoplasmic) comprises part of a switch complex (FliG, FliM, and FliN) that controls the direction of flagellar rotation. These components are also implicated in torque generation. The interaction of CheY-P with FliM stabilizes the state in which the filament, viewed along its helical axis looking toward the cell, spins CW. At room temperature, the null state is CCW.

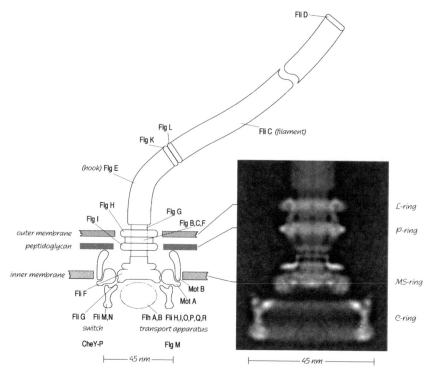

FIGURE 9.3. A schematic diagram of the flagellar rotary motor, drawn to scale. Inset: Rotationally averaged reconstruction of electron micrographs of purified hook-basal bodies. Compare Table A.3. The signaling molecule CheY-P, which binds FliM, is shown at the lower left. FlgM (lower right) blocks the activity of a sigma-factor that activates late genes. FlgM is pumped out of the cell via the transport apparatus once the basal part of the motor is complete. (Image reconstruction courtesy of David DeRosier, Brandeis University.)

It is not clear to me why the apparatus that controls the direction of rotation is called a switch. I think of a switch as something that turns an electric motor on and off, although one could have a switch that changes the sign of the current flow in its windings and, thus, its direction of rotation. Gear shift might be more appropriate, but there are no gears or transmission. However, the flagellar motor is driven by an electric current: in *E. coli*, this is a flow of protons down an electrochemical gradient, from the outside to the inside of the cell. In marine bacteria or certain bacteria that live at high pH (where protons are scarce), it is a flow of sodium ions. It is thought that protons travel from the periplasm to the C-ring/MS-ring complex via a channel that crosses the cytoplasmic membrane, mostly in MotA, causing the cytoplasmic part of MotA to pull on FliG at the periphery of the MS-ring. MotB links MotA to the peptidoglycan layer, that is, to the rigid framework of the cell wall. If this view is correct, then the C-ring/MS-ring complex serves as the rotor and MotA/MotB serves as the stator. More will be said about this in Chapter 12.

Flagellar Filament

Flagellar filaments are polymers of identical subunits, molecules of FliC, also called flagellin (named by Astbury et al., 1955). This protein can be obtained in monomeric form by removing filaments from cells mechanically, suspending them in physiological saline, and heating to 60°C. And filaments can be reconstituted from such a solution (Abram and Koffler, 1964; Asakura et al., 1964). The flagellum was recognized as an organelle of locomotion early on (see Chapter 2). Its filament scatters enough light to be seen in the light microscope (Reichert, 1909) and is readily resolved in the electron microscope (Piekarski and Ruska, 1939). Recall the electron micrograph of Fig. 2.5, and the fluorescence images of Figs. 5.4 and 5.5. Originally, the filament was thought to be a kind of primitive muscle, either a bending machine or a device that could propagate spiral waves. Later, it proved to be simply a propeller (Berg and Anderson, 1973; Silverman and Simon, 1974). The flagellin subunits are arranged on the surface of a cylinder in two different ways, as illustrated in Fig. 5.3. Their arrangement is hexagonal, with 1-, 5-, 6-, and 11-start helices. The 11-start helices appear as protofilaments that are nearly longitudinal. As discussed in Chapter 5, the subunits in one form (L-type, left) are farther apart, and in the other form (R-type, right) they are closer together. If filaments are constructed of only one type of protofilament, as shown in the figure, they are straight, with a left-handed or right-handed twist, respectively. If they are constructed of both types of protofilament, they are helical, with curvature as well as twist. On the assumption that the elastic strain energy is minimized when protofilaments of the same type are adjacent to one another, 12 different forms are predicted (two straight and 10 helical, with 1, 2, ..., 9, or 10 protofilaments in the R form, respectively; see Calladine, 1978). For the helices shown in Fig. 5.2, 2, 4, 5, or 6 of the protofilaments are in the R form, respectively. In solution, the flagellin molecule is disordered at both its N- and C-termini. The ends of the molecule become ordered as subunits polymerize, forming α-helical coiled coils in two cylindrical shells near the core of the filament, surrounding a 3 nm pore (see Namba and Vonderviszt, 1997). The central part of the flagellin molecule ends up on the outside of the filament and tolerates large structural modification. A truncated form of flagellin, formed by clipping off peptides from either end of the molecule, has been crystallized, yielding a structure for the R-type subunit. When this structure is stretched via computer simulation, it snaps into a putative L-type form (Samatey et al., 2001). Complete atomic models of both the R-type and L-type filaments should be available soon.

References

Abram, D., and H. Koffler. 1964. In vitro formation of flagella-like filaments and other structures from flagellin. *J. Mol. Biol.* 9:168–185.

Adler, J. 1969. Chemoreceptors in bacteria. *Science* 166:1588–1597.

Asakura, S., G. Eguchi, and T. Iino. 1964. Reconstitution of bacterial flagella in vitro. *J. Mol. Biol.* 10:42–56.

Astbury, W. T., E. Beighton, and C. Weibull. 1955. The structure of bacterial flagella. *Symp. Soc. Exp. Biol.* 9:282–305.

Berg, H. C., and R. A. Anderson. 1973. Bacteria swim by rotating their flagellar filaments. *Nature* 245:380–382.

Calladine, C. R. 1978. Change in waveform in bacterial flagella: the role of mechanics at the molecular level. *J. Mol. Biol.* 118:457–479.

Falke, J. J., R. B. Bass, S. L. Butler, S. A. Chervitz, and M. A. Danielson. 1997. The two-component signaling pathway of bacterial chemotaxis: a molecular view of signal transduction by receptors, kinases, and adaptation enzymes. *Annu. Rev. Cell Dev. Biol.* 13:457–512.

Falke, J. J., and G. L. Hazelbauer. 2001. Transmembrane signaling in bacterial chemoreceptors. *Trends Biochem. Sci.* 26:257–265.

Iino, T., Y. Komeda, K. Kutsukake, et al. 1988. New unified nomenclature for the flagellar genes of *Escherichia coli* and *Salmonella typhimurium*. *Microbiol. Rev.* 52:533–535.

Kim, K. K., H. Yokota, and S.-H. Kim. 1999. Four helical bundle structure of the cytoplasmic domain of a serine chemotaxis receptor. *Nature* 400:787–792.

Kim, S.-H., W. Wang, and K. K. Kim. 2002. Dynamic and clustering model of bacterial chemotaxis receptors: structural basis for signaling and high sensitivity. *Proc. Natl. Acad. Sci. USA* 99:11611–11615.

Lee, S.-Y., H. S. Cho, J. G. Pelton, et al. 2001. Crystal structure of an activated response regulator bound to its target. *Nature Struct. Biol.* 8:52–56.

Namba, K., and F. Vonderviszt. 1997. Molecular architecture of bacterial flagellum. *Q. Rev. Biophys.* 30:1–65.

Parkinson, J. S., and E. C. Kofoid. 1992. Communication modules in bacterial signaling proteins. *Annu. Rev. Genet.* 26:71–112.

Piekarski, G., and H. Ruska. 1939. Übermikroskopische Darstellung von Bakteriengeisseln. *Klin. Wochenschr.* 18:383–386.

Reichert, K. 1909. Über die Sichtbarmachung der Geisseln und die Geisselbewegung der Bakterien. *Zentralbl. Bakteriol. Parasitenk. Infektionskr. Abt. 1 Orig.* 51:14–94.

Samatey, F. A., K. Imada, S. Nagashima, et al. 2001. Structure of the bacterial flagellar protofilament and implications for a switch for supercoiling. *Nature* 410:331–337.

Silverman, M., and M. Simon. 1974. Flagellar rotation and the mechanism of bacterial motility. *Nature* 249:73–74.

Taylor, B. L., I. B. Zhulin, and M. S. Johnson. 1999. Aerotaxis and other energy-sensing behavior in bacteria. *Annu. Rev. Microbiol.* 53:103–128.

10
Genetics and Assembly

Genetic Map

Some 50 genes encode products necessary for the assembly and operation of the chemotaxis system. These are shown on the genetic map of *E. coli* in Fig. 10.1. Arrows indicate operon structure, as described in the figure legend. Most of these genes specify components required for construction of the flagellar rotary motor. They fall into three hierarchical sets (Table 10.1). The early set specifies the transcriptional regulators, *FlhD* and *FlhC*, required for expression of all the other genes. The middle set encodes components of the hook-basal body, including the transport apparatus, rotor, drive shaft, bushing, hook, hook-associated proteins, and filament cap; recall Fig. 9.3. It also encodes a protein (FliA, alias σ^F or σ^{28}) destined to turn on the late genes, together with a protein, FlgM, that inactivates it. The regulatory proteins are listed in Table A.4 in the appendix. FlgM is pumped out of the cell by the transport apparatus when the hook-basal body is complete (Hughes et al., 1993; Kutsukake, 1994). This allows expression of genes that encode the filament (FliC), the force generators (MotA, MotB), and everything else required for direction control (receptors and *che*-gene products).

Essentially all of the genes and gene products required for bacterial chemotaxis are now known. Possible exceptions are genes (and gene products) required for essential cellular functions. Since a defect in such a gene blocks cell growth (is lethal), a more subtle approach is needed to learn whether it might be required for chemotaxis. The method of choice is isolation of conditional-lethal mutants, for example, mutants that are wild-type at one temperature but mutant at another temperature. Can cells be found that when grown at the permissive temperature and then switched to the nonpermissive one become nonchemotactic? Relatively little work of this kind has been done, because most, if not all, of the

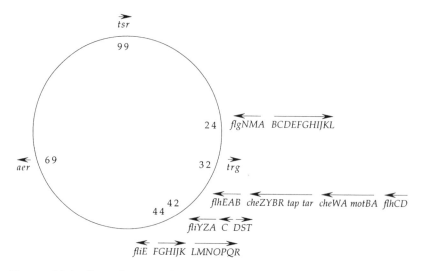

FIGURE 10.1. Genetic map of *E. coli* showing locations of the genes that make up the flagellar regulon. Arrows indicate operons (groups of genes transcribed on a single messenger RNA) and the direction of transcription. Transcription is initiated at a promoter, activated by a transcription factor; see Table 10.1. The genome is a closed circle of double-stranded DNA. It is calibrated in minutes, from 0 to 100, starting at the top. This scale arose from the time intervals required for the transfer of DNA from one cell to another during bacterial mating. The genes for many of the components of the basal body are located near 24 minutes; most of the others are between 42 and 44 minutes.

TABLE 10.1. Operon hierarchy for genes encoding proteins of *E. coli*'s chemotaxis system.

Early genes	Middle genes	Late genes
flhDC	*flgAMN*	*fliC*
	flgBCDEFGHIJKL	*motABcheAW*
	flhBAE	*tar tap cheRBYZ*
	fliAZY	*aer*
	fliDST	*trg*
	fliE	*tsr*
	fliFGHIJK	
	fliLMNOPQR	

Note: The genes that are underlined belong to the operons shown, activated by FlhDC, but they have additional promoters activated by FliA. Thus, they are expressed partially as middle genes and fully as late genes.

genes required for bacterial chemotaxis are not essential. Indeed, *E. coli* turns off the transcription (the reading) of its chemotaxis genes when grown in a suitably rich environment. And many laboratory strains of *E. coli* are nonmotile, because they have not had to compete for scarce nutrients. In short, if there is no need to hunt for food, then why bother to build the chemotaxis apparatus? Or if you try to make motors and something goes wrong, don't waste time and energy making filaments (an enormous undertaking).

The downregulation of the *flhDC* operon that results from growth on glucose (called catabolite repression) is well understood. It involves a cyclic adenosine monophosphate (cAMP) binding protein that interacts with DNA at a site upstream of *flhDC*. This protein activates *flhDC* and a number of other operons, but only when cAMP is present. The synthesis of cAMP is suppressed by a glucose catabolite. The upregulation of the *flhDC* operon that results from growth on a rich medium near a solid surface that leads to the hyperflagellation required for swarming (see Chapter 3) is poorly understood. *flhDC* is called the master operon. It is the target of a number of signals that gauge the state of the cell cycle and the external environment.

Flagellar Assembly

The motor is built from the inside out. Copies of FliF form the MS-ring, FliG,M,N the switch complex, and FlhA,B FliH,I,O,P,Q,R the transport apparatus (used for export of axial motor components); again, recall Fig. 9.3. Then copies of FlgB,C,F,G form the rod, FlgI the P-ring, FlgH the L-ring, and FlgE the hook. FliH and I are exported through the inner membrane by a different transport mechanism that involves cleavage of an N-terminal signal peptide. FliK is involved in switching export from FlgE to the hook-associated proteins, FlgK,L and FliD, and to the filament protein, FliC. FlgK,L and FliD assemble at the distal end of the hook, and FliC polymerizes under the FliD cap. The filament grows at its distal, not its proximal, end. Therefore, the flagellin subunits must pass through its axial pore. The FliD cap is essential for filament growth. If the cap is missing, flagellin simply leaks out into the external medium. A cap rotation mechanism promotes filament assembly (Yonekura et al., 2000). Finally, MotA,B appear at the periphery of the MS-ring/switch complex, and the cell becomes motile.

Other components play accessory roles. FlgJ is a muramidase that cuts a hole through the peptidoglycan for the elongating rod, FlgA is a chaperone that assists in the assembly of FlgI into the P-ring, FlgD is a hook capping protein, and FlgN and FliS, T are chaperones that keep hook-associated and filament proteins unfolded until successfully transported. Control also occurs post-transcriptionally. For example, translation of the messenger RNA for the hook protein, FlgE, is suppressed after the transport apparatus is assembled but before the construction of the rod is complete. Mechanisms of this kind can regulate the synthesis of different proteins encoded at the same level of the transcriptional hierarchy.

The late genes also encode components required for control of the direction of flagellar rotation, that is, for chemoreception and signal processing (see Chapter 9).

For reviews on flagellar assembly and the control of flagllar gene expression, see Aizawa (1996), Chilcott and Hughes (2000), Aldridge and Hughes (2002), and Macnab (2003).

References

Aizawa, S.-I. 1996. Flagellar assembly in *Salmonella typhimurium*. *Mol. Microbiol.* 19:1–5.

Aldridge, P., and K. T. Hughes. 2002. Regulation of flagellar assembly. *Curr. Opin. Microbiol.* 5:160–165

Chilcott, G. S., and K. T. Hughes. 2000. Coupling of flagellar gene expression to flagellar assembly in *Salmonella enterica* serovar typhimurium and *Escherichia coli*. *Microbiol. Mol. Biol. Rev.* 64:694–708.

Hughes, K. T., K. L. Gillen, M. J. Semon, and J. E. Karlinsey. 1993. Sensing structural intermediates in bacterial flagellar assembly by export of a negative regulator. *Science* 262:1277–1280.

Kutsukake, K. 1994. Excretion of the anti-sigma factor through a flagellar substructure couples flagellar gene expression with flagellar assembly in *Salmonella typhimurium*. *Mol. Gen. Genet.* 243:605–612.

Macnab, R. M. 2003. How bacteria assemble flagella. *Annu. Rev. Microbiol.* 57:77–100.

Yonekura, K., S. Maki, D. G. Morgan, et al. 2000. The bacterial flagellar cap as the rotary promoter of flagellin self-assembly. *Science* 290: 2148–2152.

11
Gain Paradox

Receptor Sensitivity

Data obtained early on suggested that the chemotactic response is proportional to the change in receptor occupancy, with that occupancy characterized by a fixed dissociation constant, K_d, the concentration of ligand at which the probability of receptor occupancy is 1/2 (Berg and Tedesco, 1975; Mesibov et al., 1973). Then it became evident that the dissociation constant increases (i.e., cells become less sensitive) at higher concentrations of ligand, as receptors are methylated (Borkovich et al., 1992; Bornhorst and Falke, 2000; Dunten and Koshland 1991; Li and Weis, 2000). However, even at these higher concentrations (e.g., for the non-metabolizable aspartate analog α-methylaspartate at an ambient concentration of 0.16 mM) the gain is prodigious: a step increase in concentration from 0.16 to 0.16 + 0.0027 mM (a change of about 1.7%) transiently increases the probability that the motor spins counterclockwise (CCW) by 0.23 (Segall et al., 1986). Computer simulations of the chemotaxis system (e.g., Bray et al., 1993; reviewed by Bray, 2002) fail to predict the necessary gain. Two recent findings appear to resolve the paradox. First, there is an amplification step at the beginning of the signaling pathway: the fractional change in kinase activity is some 35 times larger than the fractional change in receptor occupancy (Sourjik and Berg, 2002a). Second, there is another amplification step at the end of the signaling pathway: the motor is ultrasensitive (Cluzel et al., 2000); see below.

Evidence for the first amplification step was obtained by using fluorescence resonance energy transfer (FRET) to monitor the interaction between the response regulator, CheY-P, and its phosphatase, CheZ. At steady state, the concentration of CheY-P is constant: CheY is phosphorylated at the same rate that it is dephosphorylated. The dephosphorylation rate is proportional to the concentration of the CheY-P/CheZ complex, so from that concentration one can deduce the relative activity of the kinase. The receptor occupancy can be estimated from values for the K_d measured in vitro. One makes a fusion protein between cyan fluorescent protein (CFP) and CheZ, and another fusion protein between yellow fluorescent protein (YFP) and CheY, excites the CFP, and measures the fluorescence emission from both CFP and YFP. If the fluorophores of CFP and YFP are closer to one another than about 10 nm, which is the case for the CheY-P/CheZ complex, energy is transferred from CFP to YFP. As a result, the CFP emis-

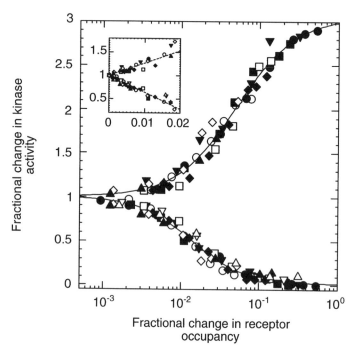

FIGURE 11.1. Fractional change in the activity of the kinase, CheA, upon addition and removal of a nonmetabolizable aspartate analog, α-methylaspartate. The initial activity is 1, and it falls to zero upon addition of enough attractant to saturate the response (lower curve). Given time (several minutes) the cells adapt, and the activity returns to 1. Then, when the attractant is removed, the activity increases (upper curve). These experiments were done with different ambient concentrations of α-methylaspartate, ranging from 0 (closed circles) to 10 mM (open triangles). The insert shows the same data plotted on a linear scale. (From Sourjik and Berg, 2002a, Fig. 3C.)

sion goes down and the YFP emission goes up. Results from this kind of analysis are shown is Fig. 11.1. Data obtained over a wide range of ambient concentrations, indicated by the different symbols, collapse into a single set of curves. The inset shows these data plotted with a linear abscissa. The slopes of these plots are not ± 1, as expected, but about ± 35. The change in receptor occupancy that occurs during chemotaxis is relatively small. Cells swimming up spatial gradients of aspartate operate near the left end of the lower curve: substantial extensions of run length occur for fractional changes in receptor occupancy as small as 0.002. How is this amplification achieved?

Receptor Clustering

The answer appears to be via receptor–receptor interactions. Receptors tend to cluster, usually at one pole, as shown in the electron microscope by immunogold labeling (Maddock and Shapiro, 1993). Clustering also is evident from the distribution of fluorescent fusion proteins (Fig. 11.2). In addition to receptors, the clusters include CheA, CheB, CheR, CheW, CheY, and CheZ. CheA and CheW bind to the receptor signaling domain, CheB and CheY to CheA, CheR to the receptor C-terminal peptide, and CheZ to the short form of CheA, $CheA_S$. If either CheA or CheW are missing, receptors still appear at the poles, but as diffuse caps, and the other components normally associated with clusters (except CheR) spread throughout the cytoplasm.

The presence of receptor clusters at one pole led to the suggestion that E. coli has a nose. However, when a cell tumbles and chooses a new run direction, either end goes first (Berg and Turner, 1995). If there is some reason for clustering, it does not have to do with how ligands in the external medium interact with the cell body, since the best that one can do is to disperse the receptors over the cell surface, and thus increase the size of the detector (Berg and Purcell, 1977). An alternative is to put receptors in clusters so that they can activate one another, and hence improve sensitivity, as argued by Duke and Bray (1999). Molecular models have been constructed to show what these clusters might look like (Shimizu et al., 2000). There is now direct genetic evidence that defects in a receptor of one type, for example, Tsr (in a region of receptor–receptor contact, identified by x-ray crystallography) can be cured by interaction with a receptor of

FIGURE 11.2. Images of cells expressing a fluorescent fusion protein, YFP-CheR. CheR, the methyltransferase, binds to receptors at their C-terminal pentapeptide, as shown in Fig. 9.2. It does this whether or not the receptors are clustered. The cells at the left are wild-type and show receptor clusters at their poles as diffraction-limited spots. The cells at the right are missing the CheA kinase, and their receptors appear, instead, as diffuse polar caps. That the clusters and caps contain receptors has been verified by labeling with anti-Tsr rabbit antibody (images not shown). Photographs courtesy of Victor Sourjik.

another type, for example, Tar (Ames et al., 2002). In addition, response to a give attractant (e.g., serine) can be enhanced by de novo receptor clustering, forced by the addition of a chemical bearing multiple copies of a different ligand (e.g., galactose) that is sensed by a different receptor (Gestwicki and Kiessling, 2002). Other evidence for cooperativity between receptors is reviewed by Falke (2002). But precisely how receptors activate one another remains to be determined.

Motor Response

The other amplification step comes from the highly cooperative response of the motor to changes in the concentration of CheY-P. The concentration of CheY-P (actually, of a fusion between CheY-P and green fluorescent protein) was measured in single cells by fluorescence correlation spectroscopy. Every cell behaved identically: a shift in concentration of CheY-P from 2.7 to 3.5 μM was enough to change the probability of clockwise (CW) rotation from 0.2 to 0.8 (Cluzel et al., 2000). These data are shown in Fig. 11.3, along with data for the binding of CheY-P to FliM obtained by FRET (Sourjik and Berg, 2002b). The binding curves are not as good as they might be, because a substantial fraction of CFP-FliM was free in the cytoplasm. But both sets of data can be fit by the two-state allosteric model of Monod et al. (1965). A more general allosteric model for motor switching has been developed by Duke et al. (2001) in which FliM molecules in a ring of 34 bind CheY-P and interact with their neighbors. Each protein can adopt a CW or a CCW configuration, and the direction of rotation depends on how many proteins are in either state. Given a large-enough interaction energy between adjacent molecules, the ensemble switches from a state in which nearly all are in the CW configuration to one in which nearly all are in the CCW configuration. But once again, the details of the mechanism remain to be determined.

Precise Adaptation

For the system to operate on such a steep response curve (Fig. 11.3), the adaptation mechanism must be precise. Is this accomplished by an as yet unknown feedback mechanism, or is adaptation intrinsically exact (Barkai and Leibler, 1997)? Under the conditions of the tracking experiments (see Chapter 4), it was found that adaptation to aspartate was exact, while that to serine was not: the mean run length in 1 mM serine was about three times longer than the mean run length in the absence of serine. However, cells drifted up gradients of either attractant perfectly well. So adaptation need not be exact, but it has to be sufficiently precise to keep cells somewhere near the middle of the motor response curve. In the model of Barkai and Leibler, receptors are in either of two states: active or inactive. In one embodiment, perfect adaptation is achieved by allowing only methylated receptors to be active, specifying that CheR works at saturation, and letting CheB-P act only on receptors that are active. In this scheme, adaptation is robust, in the sense that return to the initial steady state occurs even when the concentrations of system components vary widely. This proposition has been confirmed experimentally (Alon et al., 1999).

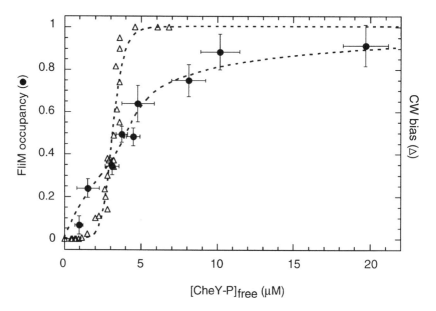

FIGURE 11.3. Comparison of the dependence of motor bias (△) and FliM occupancy (●) on concentration of free cytoplasmic CheY-P. Data for the motor bias are from Cluzel et al. (2000) and for the FliM binding from Sourjik and Berg (2002b). Dashed lines are fits to an allosteric model, showing a highly cooperative state function and a nearly linear binding function; see the text. (Adapted from Sourjik and Berg, 2002b, Fig. 2b.)

A Modelers' Era

We are entering a new phase in the study of chemotaxis in which enough is known about the detailed properties of the signaling network that its behavior can be treated analytically and simulated numerically. It is becoming a subject more in tune with the physical sciences, with constructive interplay between theory and experiment.

References

Alon, U., M. G. Surette, N. Barkai, and S. Leibler. 1999. Robustness in bacterial chemotaxis. *Nature* 397:168–171.

Ames, P., C. A. Studdert, R. H. Reiser, and J. S. Parkinson. 2002. Collaborative signaling by mixed chemoreceptor teams in *Escherichia coli*. *Proc. Natl. Acad. Sci. USA* 99:7060–7065.

Barkai, N., and S. Leibler. 1997. Robustness in simple biochemical networks. *Nature* 387:913–917.

Berg, H. C., and E. M. Purcell. 1977. Physics of chemoreception. *Biophys. J.* 20:193–219.

Berg, H. C., and P. M. Tedesco. 1975. Transient response to chemotactic stimuli in *Escherichia coli. Proc. Natl. Acad. Sci. USA* 72:3235–3239.

Berg, H. C., and L. Turner. 1995. Cells of *Escherichia coli* swim either end forward. *Proc. Natl. Acad. Sci. USA* 92:477–479.

Borkovich, K. A., L. A. Alex, and M. I. Simon. 1992. Attenuation of sensory receptor signaling by covalent modification. *Proc. Natl. Acad. Sci. USA* 89:6756–6760.

Bornhorst, J. A., and J. J. Falke. 2000. Attractant regulation of the aspartate receptor-kinase complex: limited cooperative interactions between receptors and effects of the receptor modification state. *Biochemistry* 39:9486–9493.

Bray, D. 2002. Bacterial chemotaxis and the question of gain. *Proc. Natl. Acad. Sci. USA* 99:7–9.

Bray, D., R. B. Bourret, and M. I. Simon. 1993. Computer simulation of the phosphorylation cascade controlling bacterial chemotaxis. *Mol. Biol. Cell* 4:469–482.

Cluzel, P., M. Surette, and S. Leibler. 2000. An ultrasensitive bacterial motor revealed by monitoring signaling proteins in single cells. *Science* 287:1652–1655.

Duke, T. A. J., and D. Bray. 1999. Heightened sensitivity of a lattice of membrane receptors. *Proc. Natl. Acad. Sci. USA* 96:10104–10108.

Duke, T. A. J., N. Le Novère, and D. Bray. 2001. Conformational spread in a ring of proteins: a stochastic approach to allostery. *J. Mol. Biol.* 308:541–553.

Dunten, P., and D. E. Koshland, Jr. 1991. Tuning the responsiveness of a sensory receptor via covalent modification. *J. Biol. Chem.* 266:1491–1496.

Falke, J. L. 2002. Cooperactivity between bacterial chemotaxis receptors. *Proc. Natl. Acad. Sci. USA* 99:6530–6532.

Gestwicki, J. E., and L. L. Kiessling. 2002. Inter-receptor communication through arrays of bacterial chemoreceptors. *Nature* 415:81–84.

Li, G., and R. M. Weis. 2000. Covalent modification regulates ligand binding to receptor complexes in the chemosensory system of *Escherichia coli. Cell* 100:357–365.

Maddock, J. R., and L. Shapiro. 1993. Polar location of the chemoreceptor complex in the *Escherichia coli* cell. *Science* 259:1717–1723.

Mesibov, R., G. W. Ordal, and J. Adler. 1973. The range of attractant concentrations for bacterial chemotaxis and the threshold and size of response over this range. *J. Gen. Physiol.* 62:203–223.

Monod, J., J. Wyman, and J.-P. Changeux. 1965. On the nature of allosteric transitions: a plausible model. *J. Mol. Biol.* 12:88–118.

Segall, J. E., S. M. Block, and H. C. Berg. 1986. Temporal comparisons in bacterial chemotaxis. *Proc. Natl. Acad. Sci. USA* 83:8987–8991.

Shimizu, T. S., N. Le Novère, M. D. Levin, A. J. Beavil, B. J. Sutton, and D. Bray. 2000. Molecular model of a lattice of signalling proteins involved in bacterial chemotaxis. *Nature Cell Biol.* 2:1–5.

Sourjik, V., and H. C. Berg. 2000. Localization of components of the chemotaxis machinery of *Escherichia coli* using fluorescent protein fusions. *Mol. Microbiol.* 37:740–751.

Sourjik, V., and H. C. Berg. 2002a. Receptor sensitivity in bacterial chemotaxis. *Proc. Natl. Acad. Sci. USA* 99:123–127.

Sourjik, V., and H. C. Berg. 2002b. Binding of the *Escherichia coli* response regulator CheY to its target measured in vivo by fluorescence resonance energy transfer. *Proc. Natl. Acad. Sci. USA* 99:12669–12674.

12
Rotary Motor

The structure of the rotary motor was described in Chapter 9 (Fig. 9.3) and its assembly was discussed in Chapter 10. Here, I will say more about function. Given that the diameter of the motor is less than one-tenth the wavelength of light and that it contains more than 20 of different kinds of parts (Appendix, Table A.3), it is a nanotechnologist's dream (or nightmare).

Power Source

Flagellar motors of *E. coli* are not powered by adenosine triphosphate (ATP) the fuel that energizes muscles (Larsen et al., 1974), but rather by protons moving down an electrochemical gradient; other cations and anions have been ruled out (Ravid and Eisenbach, 1984). The work per unit charge that a proton can do in crossing the cytoplasmic membrane is called the protonmotive force, Δp. In general, it comprises two terms, one due to the transmembrane electrical potential difference, $\Delta \psi$, and the other to the transmembrane pH difference $(-2.3\,kT/e)\,\Delta pH$, where k is Boltzmann's constant, T the absolute temperature, and e the proton charge. At 24°C, $2.3\,kT/e = 59\,\text{mV}$. By convention, $\Delta \psi$ is the internal potential less the external potential, and ΔpH is the internal pH less the external pH. *E. coli* maintains its internal pH in the range 7.6 to 7.8. For cells grown at pH 7, $\Delta p \approx -170\,\text{mV}$, $\Delta \psi \approx -120\,\text{mV}$, and $-59\,\Delta pH \approx -50\,\text{mV}$. For cells grown at pH 7.7, $\Delta p \approx \Delta \psi \approx -140\,\text{mV}$. For a general discussion of chemiosmotic energy coupling, see Harold and Maloney (1996).

The dependence of speed on voltage has been measured in *E. coli* by wiring motors to an external voltage source. Filamentous cells were drawn roughly halfway into micropipettes, and the cytoplasmic membrane of the segment of the cell inside the pipette was made permeable to ions by exposure to the ionophore gramicidin S. An inert marker was attached to a flagellar motor on the segment of the cell outside the pipette, and its motion was recorded on videotape. Application of an electrical potential between the external medium and the inside of the pipette (the latter negative) caused the marker to spin (Fung and Berg, 1995). The rotation speed was directly proportional to Δp over the full physiological range (up to $-150\,mV$). These experiments were done with large markers (heavy loads) at speeds less than $10\,Hz$. They have been repeated in a different way with small markers (light loads) at speeds up to nearly $300\,Hz$, and the rotation speed still appears proportional to Δp (Gabel and Berg, 2003).

The only measurement of proton flux that has been made is with motors of the motile *Streptococcus* sp. strain V4051 (van der Drift et al., 1975), a peritrichously flagellated, primarily fermentative, gram-positive organism that lacks an endogenous energy reserve and is sensitive to ionophores and uncouplers. Unlike *E. coli*, this organism can be starved and artificially energized, either with a potassium diffusion potential (by treating cells with valinomycin and shifting them to a medium with a lower concentration of potassium ion) or with a pH gradient (by shifting cells to a medium of lower pH). If this is done with a medium of low buffering capacity, one can follow proton uptake by the increase in external pH. The frequency of rotation of filaments in flagellar bundles can be determined by using power spectral analysis to measure cell vibration frequencies (Lowe et al., 1987). Finally, the data can be normalized to single motors by counting the number of cells and the number of flagellar filaments per cell. The total proton flux into the cell is much larger than the flux through its flagellar motors. However, the two can be distinguished by suddenly stopping the motors by adding an antifilament antibody—this cross-links adjacent filaments in the flagellar bundles—and measuring the change in flux. This change was found to be directly proportional to the initial swimming speed, as would be expected if a fixed number of protons carries a motor through each revolution. This number is about 1200 (Meister et al., 1987) but subject to uncertainty, due mainly to the difficulty of counting flagellar filaments.

Some bacteria, notably marine bacteria or bacteria that live at high pH, use sodium ions instead of protons (Imae, 1991; Imae and Atsumi, 1989). Thus, flagellar motors are ion driven, not just proton driven. For reviews on sodium-driven motors, see McCarter (2001) and Yorimitsu and Homma (2001).

Torque-Generating Units

The flux through the flagellar motor is divided into as many as eight distinct proton channels (or pairs of proton channels), comprising one or more copies of the proteins MotA and MotB (currently thought to be four MotA and two MotB). Evidence for this was obtained by restoring the motility of paralyzed cells (*mot* mutants) via the expression of wild-type genes carried by plasmids. As new protein is synthesized, the speed of tethered cells increases in a number of equally spaced steps, as shown in Fig. 12.1. This indicates that each additional torque-generating unit (comprising MotA and MotB) adds the same increment of torque (applies a similar force at the same distance from the axis of rotation). The main argument for a complement of eight such torque-generating units is that resurrections of this kind have produced eight equally spaced levels more than once, but never nine.

Stepping

It is likely that the passage of each proton (or each proton pair) moves a torque generator (a MotA, MotB complex) one step (one binding site) along the periphery of the rotor, suddenly stretching the components that link that generator to the rigid framework of the cell wall. As this linkage relaxes, a tethered cell should rotate by a fixed increment, once the tether relaxes (see below). In other words, the motor should behave like a stepping motor. Since proton passage is likely to occur at random times, the steps will occur with exponentially distributed waiting times. We have been looking for such steps since 1976 (Berg, 1976) but without success. The main reason, advanced then, is that the torque applied to the structure linking the rotor to the tethering surface (a series of elastic elements, comprising the rod, hook, and filament) causes that structure to twist. When less torque is applied, these elements tend to untwist, carrying the cell body forward. Therefore, discontinuities in the relative motion of rotor and stator are smoothed out. To succeed, one probably needs to work at reduced

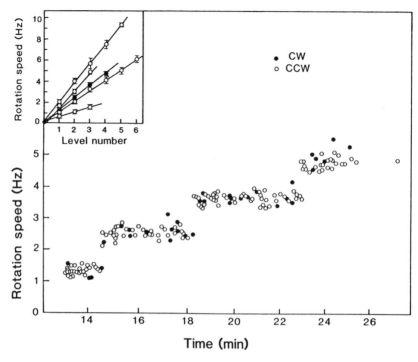

FIGURE 12.1. Rotation speed of a tethered *motA* cell, *E. coli* strain MS5037(pDFB36), following addition (at time 0) of the inducer isopropyl β-D-thiogalactoside (IPTG, added in a minimal medium containing glycerol, glucose, and essential amino acids). Filled circles indicate CW rotation, open circles CCW rotation. The inset shows the mean rotation speed ($\pm$standard error of the mean) at each level (step of the staircase) as a function of level number, for this cell (closed circles) and for four additional cells (open circles). (Blair & Berg, 1988, Fig. 1, reprinted with permission from the American Association for the Advancement of Science.)

torque, for example, with a one-generator motor driving a small viscous load, perhaps just a hook. Such an object is expected to spin quite rapidly, so the technical problems are formidable.

One route around this difficulty is to examine variations in rotation period. If n steps occur at random each revolution, then the ratio of the standard deviation to the mean should be $n^{-1/2}$ (see the appendix in Samuel and Berg, 1995), so one can determine n. With tethered wild-type cells, the answer turns out to be about 400. This work also showed that tethered cells are not free to execute rotational brownian motion. Thus, the rotor and stator are interconnected most of the time.

This stochastic analysis was repeated with tethered cells undergoing resurrection (as in Fig. 12.1), and the number of steps per revolution was found to increase linearly with level number, increasing by about 50 steps per level (Samuel and Berg, 1996). If torque generators interact with a fixed number of binding sites on the rotor, say 50, then why is the number of steps per revolution not just 50? If m torque generators are attached to the rotor and one steps, suddenly stretching its linkage to the rigid framework of the cell wall, then when that linkage relaxes and moves the rotor, it also must stretch the linkages of the $m - 1$ torque generators that have not stepped. If $m = 2$, the net movement of the rotor is half of what it would be at $m = 1$, so the apparent step number is 100 per revolution. If $m = 8$, the apparent step number is 400 per revolution. If, on the other hand, each torque generator is detached most of the time (for most of its duty cycle), then the apparent step number would remain 50. So this experiment argues not only that each force generator steps independently of all the others, but that each remains connected to the rotor most of the time. In fact, the torque generators must be attached nearly all of the time (see below).

Torque-Speed Dependence

A crucial test of any motor model is its torque-speed dependence. Measurements of the torque generated by motors of *E. coli* have been made over a wide range of speeds, including speeds in which the motor is driven backward, with the results shown in Fig. 12.2 (thick lines). At 23°C, the torque exerted by the motor is approximately constant, all the way from negative speeds of at least −100 Hz to positive speeds of nearly 200 Hz. At higher speeds it declines approximately linearly, crossing the 0-torque line at about 300 Hz. At lower temperatures, the region of transition from constant torque to declining torque—we call this the "knee"—shifts to lower speeds, and the region of decline steepens (Berg and Turner, 1993; Chen and Berg, 2000a); the latter parts of the curves can be mapped onto one another with scaling of the speed axis.

Estimates of the torque generated in the low-speed regime range from about 2.7×10^{-11} dyn cm (2700 pN nm) to 4.6×10^{-11} dyn cm (4600 pN nm), the smaller value from estimates of the viscous

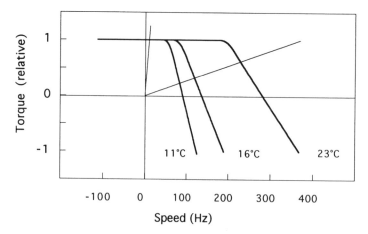

FIGURE 12.2. The torque-speed curve for the flagellar motor of *E. coli* shown at three temperatures (thick lines), together with two load lines (thin lines), one for an object the size of the cell body of wild-type *E. coli* (effective radius about 1 μm, left), the other for a latex bead of radius about 0.3 μm, right. (Adapted from Fig. 16 of Berg and Turner, 1993.) Later work showed that the torque declines somewhat in the low-speed regime, by about 10% between stall and the knee; see the text.

drag on tethered cells of *Streptococcus* (Lowe et al., 1987), and the larger value from the force exerted by tethered cells of *E. coli* on latex beads held in an optical trap (Berry and Berg, 1997).

A motor driving an inert object (a cell body, a latex bead, etc.) will spin at the speed at which the torque generated by the motor is balanced by the torque exerted on the object by viscous drag. This torque is defined by load lines, such as those shown in Fig. 12.2 (thin lines), the one at the left for a large object and the one at the right for a small object. To see this, note that the torque, N, required to rotate an object of fixed shape in a viscous medium is its rotational frictional drag coefficient, f, times its angular velocity, Ω (2π times its rotation speed, in Hz). In a torque versus speed plot, this function is a straight line passing through the origin, with slope f. Here, we assume that the medium is Newtonian, that is, that the frictional drag coefficient does not depend on Ω, a condition satisfied in a dilute aqueous medium that does not contain long unbranched molecules, such as methylcellulose or polyvinylpyrrolidone (Berg and Turner, 1979). For such a medium, f is a geometrical factor times the bulk viscosity, η, where η is independent of Ω (independent of the rate of shear). For an isolated sphere of radius a spinning about an axis through its center, for example, this geometrical factor is $8\pi a^3$. For compact globular objects, the actual shape is not very critical; however, accurate values can be computed (Garcia de la Torre and Bloomfield, 1981). The distance from the tethering surface does not really matter, either, provided that the gap between the object and the surface is at least 0.2 cell radii (Berg, 1976; Jeffery, 1915).

At 23°C and for the load line shown at the left in Fig. 12.2, the motor runs at 10 Hz; for the load line shown at the right, it runs about 220 Hz. For a very shallow load line (e.g., one for a free hook), the speed would be close to the zero-torque speed, about 290 Hz. A motor free-running in this way always operates in the upper-right-hand quadrant of Fig. 12.2. It cannot drive itself backward, although it can redefine what is meant by forward by switching from counterclockwise (CCW) to clockwise (CW) or back again. Nor can it spin faster than its speed at zero load. To probe the upper-left-hand or lower-right-hand quadrants of Fig. 12.2, one needs to subject the motor to torque applied externally.

One way to do this is by electrorotation (Washizu et al., 1993). Cells were tethered and exposed to a high frequency (2.25 MHz) rotating electric field (Berg and Turner, 1993). As explained in the latter reference, the external electric field polarizes the cell. The dipole field due to the polarization rotates at the same rate and in the same direction as the applied electric field. However, due to the finite time required for redistribution of charges, the polarization vector leads or lags the electric-field vector. The externally applied torque is the cross-product of these vectors. The applied torque varies as the square of the magnitude of the electric field and changes sign with changes in the direction of rotation of that field. Therefore, it is possible to spin a tethered cell either forward or backward. Speeds of several hundred Hz are readily attainable. For reasons that we do not understand, the motor of a cell driven backward (CW if it is trying to spin CCW, or CCW if it is trying to spin CW) often breaks catastrophically: motor torque suddenly drops to zero, the cell appears free to execute rotational brownian motion, and the motor fails to recover. Our best guess is that the C-ring is sheared off of the bottom of the rotor (Fig. 9.3), disengaging all torque-generating units but leaving the bearings intact. Once the motor has broken, one can compare the speed at which the cell body turns at a given value of externally applied torque with the speed at which it turned at the same value of externally applied torque before the break occurred. That difference is proportional to the torque generated by the motor at the speed at which it turned when intact. The data shown by the thick lines in Fig. 12.2 were determined in this way.

Additional work on the behavior of the motor in the upper-right-hand quadrant of Fig. 12.2 was done by manipulating load lines. Flagella were shortened by viscous shear, and cells were adsorbed onto positively charged glass. Latex beads of various sizes were attached to the flagellar stubs, and the slopes of their load lines were increased by addition of the viscous agent Ficoll (Chen and Berg, 2000a). In the low-speed regime, torque was found to drop by about 10% from stall to the knee. In this regime, torque was independent of temperature, and solvent isotope effects were relatively small, as found earlier for artificially energized cells of *Streptococcus* (Khan and Berg, 1983). Evidently, at low speeds, the motor operates near thermodynamic equilibrium, where rates of displacement of internal mechanical components or translocation of protons are not limiting. In the high-speed regime, torque was strongly temperature dependent, as seen in Fig. 12.2, and solvent isotope effects were large (Chen and Berg, 2000b). This is what one would expect if the decline in torque at high speed is due to limits in rates of proton transfer (proton dissociation).

Slowly declining torque in the low-speed regime argues for a model in which the rate-limiting step depends strongly on torque and dissipates most of the available free energy, that is, for a powerstroke mechanism, while the absence of a barrier to backward rotation rules out models (e.g., thermal ratchets) that contain a step that is effectively irreversible and insensitive to external torque (Berry and Berg, 1999). Eventually, we would like to understand why the low-speed regime is so broad, why the boundary between the low-speed and high-speed regimes is so narrow, and why the position of that boundary is so sensitive to temperature.

The power output, the power dissipated when a torque N sustains rotation at angular velocity Ω, is $N\Omega$. For torque 4600 pN nm and speed 10 Hz, this is 2.9×10^5 pN nm s^{-1}. The power input, the rate at which protons can do work, is proton flux times proton charge times protonmotive force. Assuming 1200 protons per revolution and speed 10 Hz, the proton flux is 1.2×10^4 s^{-1}. For *E. coli* at pH 7, $\Delta p \approx -170$ mV. Therefore, the power input is $(1.2 \times 10^4 \text{s}^{-1})$ (e) $(0.17\text{V}) = 2.0 \times 10^3$ eV s^{-1}. Since 1 eV (one electron volt) $= 1.6 \times 10^{-12}$ erg $= 160$ pN nm, the power input is 3.2×10^5 pN nm s^{-1}. So, by this crude estimate, the efficiency of the motor, power output divided by power input, is about 90%. Within the uncertainty of the measurements—the proton flux has not been measured in *E. coli*—the efficiency could be 1.

The power output, $N\Omega$, increases linearly with speed up to the boundary between the low-speed and high-speed regimes, and then it declines. If a fixed number of protons carries the motor through each revolution, the power input also increases linearly with speed. Therefore, the efficiency remains approximately constant up to the knee, and then it declines. There is no discontinuity in torque as one crosses the zero-speed axis (Berry and Berg, 1997). As the motor turns backward, it must pump protons, just as the F_0-ATPase pumps protons when driven backward by F_1.

The force exerted by each force-generating unit is substantial but not large on an absolute scale. If we take a ballpark figure for the stall torque of 4000 pN nm and assume that force-generating units act at the periphery of the rotor at a radius of about 20 nm, then 200 pN is applied. If there are eight independent force-generating units, then each contributes 25 pN. This is a force equal in magnitude to that between two electrons 4.8 angstroms apart in a medium of dielectric constant 40 (midway between water, 80, and lipid, about 2). So almost any kind of chemistry will do.

The energy available from one proton moving down the electrochemical gradient is $e\Delta p$. Given $\Delta p \approx -170$ mV, this is 0.17 eV, or 27 pN nm. At unit efficiency, this equals the work that the force-generator can do, Fd, where F is the force that it exerts, and d is the displacement generated by the transit of one proton. Assuming 52 steps per revolution (twice the number of FliG subunits) and a rotor radius of 20 nm, d ≈ 2.4 nm. So $F \approx 11$ pN. If two protons are required per elementary step, the force is twice as large, and $F \approx 22$ pN. So, given the estimate of 25 pN per force-generating unit made above, the displacement of two protons per step is likely.

Angular Dependence of Torque

When optical tweezers were used to drive cells slowly backward or to allow them to turn slowly forward (Berry & Berg, 1997), torque did not vary appreciably with angle. A very different result is obtained when one energizes and de-energizes tethered cells and asks where they stop or watches them spin when the proton-motive force is very low. When this was done with *Streptococcus*, periodicities were observed of order 5 or 6 (Khan et al., 1985). This probably reflects small periodic barriers to rotation intrinsic to the bearings.

Duty Ratio

In our stochastic analysis of steps (above) we argued that the apparent number of steps per revolution would increase with the number of torque generators, as observed, if each torque genera-tor remained attached to the rotor most of the time, that is, if the torque-generating units had a high duty ratio. The following argument shows that the duty ratio must be close to 1. Evidently, torque generators, like molecules of kinesin, are processive. Con-sider a tethered cell being driven by a single torque-generating unit, as in the first step of the resurrection shown in Fig. 12.1. If a wild-type motor with eight torque-generating units generates a torque of about 4×10^{-11} dyn cm (4000 pN nm), then the single-unit motor generates a torque of about 5×10^{-12} dyn cm. The torsional spring constant of the tether—most of the compliance is in the hook—is about 5×10^{-12} dyn cm rad^{-1} (Block et al., 1989), so the tether is twisted up about 1 radian, or 57 degrees. Now the viscous drag on the cell body is enormous compared to that on the rotor, so if the torque-generating unit lets go, the tether will unwind, driving the rotor backward. If the single unit steps 50 times per revolution, the displacement is 7.2 degrees per step. If the cell is spinning ~1.2 Hz (Fig. 12.1), the step interval is 1.6×10^{-2} s. If the duty ratio is 0.999, so that the torque-generating unit detaches for 1.6×10^{-5} s during each cycle, how far will the tether unwind? The tether unwinds exponentially: $\theta = \theta_0 \exp(-\alpha t)$, where θ_0 is the initial twist, and α is the torsional spring constant divided by the rotational frictional drag coefficient. If we approximate the rotor as a sphere of radius $a = 20$ nm immersed in a medium of viscos-ity $\eta = 1$ P (1 g cm^{-1} s^{-1}), which is about right for a lipid membrane, then the frictional drag coefficient, $8\pi\eta a^3$, is 2×10^{-16} dyn cm per rad s^{-1}, and $\alpha = 2.5 \times 10^4$ s^{-1}. So, in 1.6×10^{-5} s, the twist in the tether decreases from 57 degrees to 57 exp(-2.5×10^4 s^{-1} $\times 1.6 \times 10^{-5}$ s) = 38 degrees, or by 19 degrees, that is, by more than twice the step angle. Thus, the torque-generating unit would not be able to keep up. So the duty ratio must be close to 1. The interaction between the torque-generating unit and the rotor must be such that the rotor is not able to slip backward. If one imagines that a torque-generating unit binds to successive sites along the periphery of the rotor, then it has no unbound states. If each torque-generating unit has two proton channels (Braun and Blair, 2001), it is possible that a MotA associated with one channel remains attached to a FliG, while the MotA associated with the other channel takes the next step.

Switching

Finally, the motor can run in either direction with approximately equal efficiency. Although the force-generating elements move independently, they all switch at the same time: changes in direction occur in an all-or-none fashion within a few milliseconds. Evidently, the rotor suddenly changes shape, so that the force-generating elements step along a different track. What sort of change in conformation occurs? And why is this process so sensitive to the concentration of CheY-P?

Models

The fundamental question is how the flagellar motor generates torque, namely, how inward motion of one or more ions through a torque-generating unit causes it to advance circumferentially along the periphery of the rotor. Once that is understood, the nature of the conformational change required for switching, namely, how the direction of advance is distinguished from that of retreat, is likely to be self-evident.

Moving parts of the motor are submicroscopic and immersed in a viscous medium (water or lipid), so the Reynolds number is very small (see Chapter 6). And everything is overdamped (Howard, 2001, pp. 41–45). Thus, the designer does not have the benefit of flywheels or tuning forks. If, for example, the operator of the motor driving a tethered cell of *E. coli* 10 Hz were to put in the clutch, the cell body would coast no more than a millionth of a revolution. So if there is a stage in the rotational cycle in which the torque changes sign, the motor will stop. Predicting net torque after averaging over a complete cycle is not sufficient. And mechanisms in which energy is stored in vibrational modes are not viable. However, one can use energy available from an electrochemical potential to stretch a spring and then use that spring to apply a steady force. As we have seen, the force required is modest, and almost any kind of chemistry will do.

Motion of the torque-generating units relative to the periphery of the rotor is driven by a proton (or sodium-ion) flux. Only one experiment has attempted to measure this flux (Meister et al., 1987), and flux and speed were found to be linearly related. Unless protons flow through the motor when it is stalled, this implies that a fixed number of protons carry the motor through each revolution. The running torque at low speeds is close to the stall torque (Fig. 12.2). If the motor is stalled and no protons flow, no free energy is dissipated; therefore, the stalled motor is at thermodynamic equilibrium. For slow rotation near stall, the motor must operate reversibly at unit efficiency, with the free energy lost by protons traversing the motor equal to the mechanical work that it performs. This implies that the torque near stall should be proportional to the protonmotive force over its full physiological range, as observed. So the evidence is consistent with a model in which the motor is tightly coupled.

An important question is whether the ion that moves down the electrochemical gradient is directly involved in generating torque, that is, participates in a powerstroke in which dissipation of energy available from the electrochemical gradient and rotational work occur synchronously, or whether the ion is indirectly involved in generating torque (e.g., by enabling a ratchet that is powered thermally). In the powerstroke case, protons can be driven out of the cell by backward rotation, and steep barriers are not expected. In addition, if the rate-limiting step is strongly torque dependent, then the torque-speed curve (as plotted in Fig. 12.2) can have a relatively flat plateau, because small changes in torque can generate large changes in speed. In the ratchet case, with tight coupling, the likelihood of transit of ions against the electrochemical gradient is small, so the system must wait, even when large backward torques are applied, and barriers to backward rotation are expected. So the torque-speed curves of Fig. 12.2 favor a powerstroke mechanism.

There appear to be essential electrostatic interactions between specific residues in the cytoplasmic domain of MotA and the C-terminal domain of FliG (Zhou et al., 1998a). Here, charge complementarity is more important than surface complementarity; that is, long-range interactions appear to be more important than tight binding. Since some models for torque generation require transfer of protons from the stator to the rotor, it was expected that acidic residues on FliG might be more important than basic residues. However, replacement of the acidic residues deemed important for torque generation with alanine still allowed some rotation, while reversing their charge had a more severe effect (Lloyd and Blair, 1997). An extension of this study failed to identify any conserved basic residues critical for rotation in MotA, MotB, FliG, FliM, or FliN, and only one conserved acidic residue critical for rotation, Asp32 of MotB (Zhou et al., 1998b). Other alternatives were considered and either ruled out or deemed unlikely. Therefore, the only strong candidate for a residue that functions directly in proton conduction is Asp32 of MotB.

MotA and MotB appear to form a cassette containing a transmembrane channel that supports proton flow, generating transformations that drive movement along the periphery of the rotor. That the ion-dependence is determined solely by MotA and MotB (or their homologs) has been shown conclusively in recent experiments in which transmembrane and cytoplasmic domains of MotA and MotB were replaced by homologous parts of PomA and PomB, from *Vibrio alginolyticus*. With only the C-terminal periplasmic domain of MotB remaining, the *E. coli* motor became sodium-ion driven rather than proton driven (Asai et al., 2003).

Given the above work, I would bet on a cross-bridge mechanism of the kind that Blair and colleagues propose (Braun et al., 1999; Kojima and Blair, 2001). In such a scheme, proton transport drives a cyclic sequence in which (1) a proton binds to an outward-facing binding site; (2) the protonmotive force drives a conformational change, a powerstroke that moves the rotor forward (or stretches a spring that moves it forward) and transforms the binding site to an inward-facing site; and (3) proton dissociation triggers detachment of the cross-bridge from the rotor, its relaxation to the original shape, and reattachment to an adjacent site. If the MotA/MotB complex is two-headed, one head could remain attached while the other stepped, thus ensuring a high duty ratio.

But to be honest, we really do not understand how the motor works, i.e., how proton translocation generates torque. Modeling would help, but what is needed most is more structural information.

Reviews

For other reviews on the structure and function of proton-driven motors, see Läuger and Kleutsch (1990), Caplan and Kara-Ivanov (1993), Schuster and Khan (1994), Macnab (1996), Khan (1997), Berry and Armitage (1999), Berry (2000, 2003), Berg (2000, 2003), and Blair (2003). For a catalog of early models, see Berg and Turner (1993). For tutorials on the mathematical treatment of motor models, see Berry (2000) and Bustamante et al. (2001). The material in this chapter was adapted from Berg (2003).

References

Asai, Y., T. Yakushi, I. Kawagishi, and M. Homma. 2003. Ion-coupling determinants of Na^+-driven and H^+-driven flagellar motors. *J. Mol. Biol.* 327:453–463.

Berg, H. C. 1976. Does the flagellar rotary motor step? In: Cell Motility, Cold Spring Harbor Conferences on Cell Proliferation. R. Goldman, T. Pollard, J. Rosenbaum, editors. Cold Spring Harbor Laboratory, Cold Spring Harbor, NY, pp. 47–56.

Berg, H. C. 2000. Constraints on models for the flagellar rotary motor. *Philos. Trans. R. Soc. Lond. B* 355:491–501.

Berg, H. C. 2003. The rotary motor of bacterial flagella. *Annu. Rev. Biochem.* 72:19–54.

Berg, H. C., and L. Turner. 1979. Movement of microorganisms in viscous environments. *Nature* 278:349–351.

Berg, H. C., and L. Turner. 1993. Torque generated by the flagellar motor of *Escherichia coli*. *Biophys. J.* 65:2201–2216.

Berry, R. B. 2000. Theories of rotary motors. *Philos. Trans. R. Soc. Lond. B* 355:503–509.

Berry, R. B. 2003. The bacterial flagellar motor. In: Molecular Motors. M. Schliwa, editor. Wiley-VCH, Weinheim, pp. 111–140.

Berry, R. B., and J. P. Armitage. 1999. The bacterial flagella motor. *Adv. Microbiol. Physiol.* 41:291–337.

Berry, R. M., and H. C. Berg. 1997. Absence of a barrier to backwards rotation of the bacterial flagellar motor demonstrated with optical tweezers. *Proc. Natl. Acad. Sci. USA* 94:14433–14437.

Berry, R. M., and H. C. Berg. 1999. Torque generated by the flagellar motor of *Escherichia coli* while driven backward. *Biophys. J.* 76:580–587.

Blair, D. F. 2003. Flagellar movement driven by proton translocation. *FEBS Lett.* 545:86–95.

Blair, D. F., and H. C. Berg. 1988. Restoration of torque in defective flagellar motors. *Science* 242:1678–1681.

Block, S. M., D. F. Blair, and H. C. Berg. 1989. Compliance of bacterial flagella measured with optical tweezers. *Nature* 338:514–517.

Braun, T. F., and Blair, D. F. 2001. Targeted disulfide cross-linking of the MotB protein of *Escherichia coli*: evidence for two H^+ channels in the stator complex. *Biochemistry* 40:13051–13059.

Braun, T. F., S. Poulson, J. B. Gully, et al. 1999. Function of proline residues of MotA in torque generation by the flagellar motor of *Escherichia coli*. *J. Bacteriol.* 181:3542–3551.

Bustamante, C., D. Keller, and G. Oster. 2001. The physics of molecular motors. *Acc. Chem. Res.* 34:412–420.

Caplan, S. R., and M. Kara-Ivanov. 1993. The bacterial flagellar motor. *Int. Rev. Cytol.* 147:97–164.

Chen, X., and H. C. Berg. 2000a. Torque-speed relationship of the flagellar rotary motor of *Escherichia coli*. *Biophys. J.* 78:1036–1041.

Chen, X., and H. C. Berg. 2000b. Solvent-isotope and pH effects on flagellar rotation in *Escherichia coli*. *Biophys. J.* 78:2280–2284.

Fung, D. C., and H. C. Berg. 1995. Powering the flagellar motor of *Escherichia coli* with an external voltage source. *Nature* 375:809–812.

Gabel, C. V., and H. C. Berg. 2003. The speed of the flagellar rotary motor of *Escherichia coli* varies linearly with protonmotive force. *Proc. Natl. Acad. Sci. USA* 100:8748–8751.

Garcia de la Torre, J., and V. A. Bloomfield. 1981. Hydrodynamic properties of complex, rigid, biological macromolecules: theory and applications. *Q. Rev. Biophys.* 14:81–139.

Harold, F. M., and P. C. Maloney. 1996. Energy transduction by ion currents. In: *Escherichia coli* and *Salmonella*: Cellular and Molecular Biology. F. C. Neidhardt, R. Curtiss, J. L. Ingraham, et al., editors. ASM Press, Washington DC, pp. 283–306.

Howard, J. 2001. Mechanics of Motor Proteins and the Cytoskeleton. Sinaur Associates, Sunderland, MA.

Imae, Y. 1991. Use of Na^+ as an alternative to H^+ in energy transduction. In: New Era of Bioenergetics. Y. Mukohata, editor. Academic Press, Tokyo, pp. 197–221.

Imae, Y., and T. Atsumi. 1989. Na^+-driven bacterial flagellar motors. *J. Bioenerg. Biomembr.* 21:705–716.

Jeffery, G. B. 1915. On the steady rotation of a solid of revolution in a viscous fluid. *Proc. Lond. Math. Soc.* 14:327–338.

Khan, S. 1997. Rotary chemiosmotic machines. *Biochim. Biophys. Acta* 1322:86–105.

Khan, S., and H. C. Berg. 1983. Isotope and thermal effects in chemiosmotic coupling to the flagellar motor of *Streptococcus*. *Cell* 32:913–919.

Khan, S., M. Meister, and H. C. Berg. 1985. Constraints on flagellar rotation. *J. Mol. Biol.* 184:645–656.

Kojima, S., and D. F. Blair 2001. Conformational change in the stator of the bacterial flagellar motor. *Biochemistry* 40:13041–13050.

Larsen, S. H., J. Adler, J. J. Gargus, and R. W. Hogg. 1974. Chemomechanical coupling without ATP: the source of energy for motility and chemotaxis in bacteria. *Proc. Natl. Acad. Sci. USA* 71:1239–1243.

Läuger, P., and B. Kleutsch. 1990. Microscopic models of the bacterial flagellar motor. *Comments Theor. Biol.* 2:99–123.

Lloyd, S. A., and D. F. Blair. 1997. Charged residues of the rotor protein FliG essential for torque generation in the flagellar motor of *Escherichia coli*. *J. Mol. Biol.* 266:733–744.

Lowe, G., M. Meister, and H. C. Berg. 1987. Rapid rotation of flagellar bundles in swimming bacteria. *Nature* 325:637–640.

Macnab, R. M. 1996. Flagella and motility. In: *Escherichia coli* and *Salmonella*: Cellular and Molecular Biology. F. C. Neidhardt, R. Curtiss, J. L. Ingraham, et al., editors. ASM Press, Washington, DC, pp. 123–145.

McCarter, L. L. 2001. Polar flagellar motility of the *Vibrionaceae*. *Microbiol. Mol. Biol. Rev.* 65:445–462.

Meister, M., G. Lowe, and H. C. Berg. 1987. The proton flux through the bacterial flagellar motor. *Cell* 49:643–650.

Ravid, S., and M. Eisenbach. 1984. Minimal requirements for rotation of bacterial flagella. *J. Bacteriol.* 158:1208–1210.

Samuel, A. D. T., and H. C. Berg. 1995. Fluctuation analysis of rotational speeds of the bacterial flagellar motor. *Proc. Natl. Acad. Sci. USA* 92: 3502–3506.

Samuel, A. D. T., and H. C. Berg. 1996. Torque-generating units of the bacterial flagellar motor step independently. *Biophys. J.* 71:918–923.

Schuster, S. C., and S. Khan. 1994. The bacterial flagellar motor. *Annu. Rev. Biophys. Biomol. Struct.* 23:509–539.

van der Drift, C., J. Duiverman, H. Bexkens, and A. Krijnen. 1975. Chemotaxis of a motile *Streptococcus* toward sugars and amino acids. *J. Bacteriol.* 124:1142–1147.

Washizu, M., Y. Kurahashi, H. Iochi, et al. 1993. Dielectrophoretic measurement of bacterial motor characteristics. *IEEE Trans. Ind. Appl.* 29:286–294.

Yorimitsu, T., and M. Homma. 2001. Na^+-driven flagellar motor of *Vibrio*. *Biochim. Biophys. Acta* 1505:82–93.

Zhou, J., S. A. Lloyd, and D. F. Blair. 1998a. Electrostatic interactions between rotor and stator in the bacterial flagellar motor. *Proc. Natl. Acad. Sci. USA* 95:6436–6441.

Zhou, J., L. L. Sharp, H. L. Tang, et al. 1998b. Function of protonatable residues in the flagellar motor of *Escherichia coli*: a critical role for Asp 32 of MotB. *J. Bacteriol.* 180:2729–2735.

13
Epilogue

What We Have Learned

I have told you some things about a free-living organism only one micron in size. It is equipped with sensors that count molecules of interest in its environment, coupled to a readout device that computes whether these counts are going up or down. The output is an intracellular signal that modulates the direction of rotation of a set of rotary engines, each turning a propeller with variable pitch. Each engine (or motor) is driven, in turn, by several force-generating elements (like pistons), powered by a transmembrane ion flux. In addition to a gear shift (labeled forward and reverse but prone to shift on its own) there is a stator, a rotor, a drive shaft, a bushing, and a universal joint.

We know a great deal about what all this machinery does for the bacterium, a fair amount about the structures of the molecular components involved (particularly those that have been crystallized), and even how the organism programs their syntheses. We know less about the precise ways in which these components function.

Levels of Amazement

Some wonder how the flagellar motor possibly could have evolved. The problem here is that we do not know about earlier states. What was the flagellar motor doing, for example, before the acquisition of the propeller (if, indeed, that was the sequence of events)? Perhaps it was winding up DNA. Or maybe it was injecting toxins into other cells as part of a program of conquest. In any event, it must have been doing something that promoted the survival of the organism. Evolution is opportunistic: it builds on components already at hand. One can not turn off the organism

121

in order to redesign it, because that means extinction. You have to modify the machinery while it is running.

A Caltech friend, John Allman, an expert on the evolution of primate brains, once marveled to me about the similarity between circuits in brains and those in a Los Angeles power plant. When he visited the power plant, he discovered a hierarchy of control devices utilizing components ranging from antique to modern (e.g., mechanical relays, vacuum tubes, transistors, integrated circuits, and computers). The reason was simple: it was desirable to improve the design without interrupting the service. In biology, this is imperative.

The flagellar motor, albeit amazing, is no more so than a number of other molecular machines. Among these are enzymes used to make RNA copies of DNA templates, that is, RNA polymerase, or macromolecular ensembles used to translate these copies into sequences of amino acids, that is, the ribosome. The latter is particularly remarkable, because it dates from an ancient era in which catalytic functions were carried out by RNA rather than protein. The structures and functions of these machines are currently being examined in atomic detail. But unless you work in a chemical plant, everyday analogs of these devices are not readily at hand. However, everyone knows about rotary motors, including those with propellers. The speed of the flagellar motor is much faster than that of the motor of a boat, something like the speed of a table saw. And if you studied Chapter 6, you will know that the physics used by the flagellar filament is rather different from that used by the propeller of a boat—it shears water rather than accelerates it. Also, the flagellar motor is very small. Richard Feynmann once offered a prize to anyone who could build a rotary motor smaller than 1/64-inch on a side. The winning model is displayed behind glass in the hallway of one of Caltech's physics buildings. The flagellar motor is more like 1/640,000 of an inch on a side! That's a million million times smaller in volume.

Where We Go from Here

Our next task is to understand a number of things more quantitatively. We are trying to develop better ways of monitoring the concentration of the signaling molecule CheY-P in living cells, with the aim of understanding more about receptor function. Why is the gain of the chemotaxis system so high, and why does adaptation work so well? As noted earlier, detailed understanding of the force-generating and switching mechanisms of the flagellar motor probably awaits crystal structures. Might it be possible to crystallize the entire machine? And more needs to be learned about the precise way in which the transport apparatus decides what components are sent along the channel leading, ultimately, from the cytoplasm to the filament cap. At the genetic level, we need to know a great deal more about the mechanisms that up- or downregulate flagellar synthesis. How, for example, does the cell decide to make many flagella and swarm over surfaces?

Motivation

Is any of this knowledge practical? The reading of the external environment by cells of all types, leading to responses in growth or motility, is fundamental to life. Bacterial chemotaxis provides a model for learning how such processes can work. However, this is not what has motivated me. I have wanted to know, simply, how such a tiny creature does its thing. How, for example, has it solved the problem of finding greener pastures within the constraints imposed by physics? This is a matter of curiosity. Curiosity is the driving force of basic science.

Appendix: Parts Lists

Parts are listed alphabetically in tables that define different functions, i.e., chemoreception (Table A.1), signal processing (Table A.2), motor output (Table A.3), and gene regulation (Table A.4). Components involved in chemoreception are found near the surface of the cell, either between the outer and inner membranes or spanning the inner membrane. Components involved in signal processing are found in the cytoplasm. Components involved in motor output are either exposed to the cytoplasm, span the inner or outer membranes, or extend out into the external medium. Components involved in gene regulation are found in the cytoplasm. One of these, FlgM, can be pumped out into the external medium.

TABLE A.1. Proteins involved in chemoreception.[a]

Gene product	Binds or senses	Size (kd)	Gene map loc. (min)
Periplasmic binding proteins[b]			
DppA	Di or tripeptides, then Tap	60	80
MalE	Maltose, then Tar	43	91
MglB	Galactose, then Trg	36	48
NikA	Ni^{2+}, then Tar	59	78
RbsB	Ribose, then Trg	31	85
Transmembrane receptors/transducers[c]			
Tap	DppA	58	42
Tar (MCP2)	Aspartate, MalE, NikA	60	42
Trg (MCP3)	MglB, RbsB	59	32
Tsr (MCP1)	Serine	59	99
Transmembrane receptors also involved in transport of sugars and sugar alcohols[d]			
BglF	β-glucosides	66	84
FruA (PtsF)	Fructose	58	49
GatA	Galacitol	17	47
GutA (SrlA)	Glucitol (sorbitol)	21	61
MtlA	Mannitol	68	81
NagE	N-acetyl glucosamine	68	15
PtsG	Glucose	51	25
PtsM	Mannose	31	41
Cytoplasmic receptors bound to the inner membrane			
Aer	Redox potential	55	69

[a] These data are for *E. coli* K12. The gene product has the same name as the gene, except it is capitalized and not italicized. Size for the gene product is given in kilodaltons (kd, thousands molecular weight). Map location for the gene is given in minutes (0–100); this calibration is based on times required for DNA transfer during bacterial mating.

[b] The periplasm is the space between the inner and outer membranes. These components also are involved in transport.

[c] These components span the inner membrane. They bind a chemical either directly or indirectly, via its binding protein. They also are called methyl-accepting chemotaxis proteins (MCPs), because they are methylated in the course of the chemotactic response.

[d] These are components EnzII of the sugar phosphotransferase system (PTS). Their substrates are phosphorylated when transported through the inner membrane.

TABLE A.2. Proteins involved in signal processing.[a]

Gene product	Function	Size (kd)	Gene map loc. (min)
Components that process signals generated by MCPs[b]			
CheA$_L$[c]	When activated by an MCP, transfers phosphate to CheB, CheY	71	42
CheB	When phosphorylated, demethylates MCPs	37	42
CheR	Methylates MCPs	33	42
CheW	Couples CheA to MCPs	18	42
CheY	When phosphorylated, binds to the motor and promotes CW rotation	14	42
CheZ	Accelerates removal of phosphate from CheY-P	24	42
Components involved in the sugar phosphotransferase system			
HPr (PtsH)	Transfers phosphate from EnzI to EnzII, or for glucose, to EnzIII	9	55
EnzI (PtsI)	Transfers phosphate from phosphoenolpyruvate (PEP) to HPr and modulates activity of CheA	64	55
EnzIII^{glc} (Crr)	Transfers phosphate from HPr to EnzII^{glc}	18	55

[a] See note a, Table A.1.
[b] See note c, Table A.1.
[c] There also is a short form of CheA, missing 97 amino acids at its N-terminus, including the phosphorylation site. The long form is called CheA$_L$, the short form CheA$_S$.

TABLE A.3. Proteins involved in motor output.

Gene product	Function or component	Size (kd)	Gene map loc. (min)
FlgA	P-ring assembly	24	24
FlgB	Proximal rod	15	24
FlgC	Proximal rod	14	24
FlgD	Hook cap	24	24
FlgE	Hook	42	24
FlgF	Proximal rod	26	24
FlgG	Distal rod	28	24
FlgH	L-ring	25	24
FlgI	P-ring	38	24
FlgJ	Muramidase	34	25
FlgK	Hook-filament junction; at hook	58	25
FlgL	Hook-filament junction; at filament	34	25
FlgN	FlgK, FlgL chaperone	16	24
FlhA[a]	Protein export	75	42
FlhB[a]	Hook-length control	42	42
FlhE	?	14	42
FliC	Filament (flagellin)	51	43
FliD	Filament cap	48	43
FliE	Rod MS-ring junction (?)	11	43
FliF	MS-ring	61	43
FliG	Switch component; interacts with MotA	37	43
FliH	Protein export	26	43
FliI[a]	Protein export ATPase	49	43
FliJ	Rod, hook, and filament chaperone	17	43
FliK	Hook-length control	39	43
FliL	Inner-membrane associated; unknown function	17	43
FliM	Switch component; binds CheY-P	38	44
FliN	Switch component	15	44
FliO	Protein export	11	44
FliP[a]	Protein export	27	44
FliQ[a]	Protein export	10	44
FliR[a]	Protein export	29	44
FliS	FliC chaperone	15	43
FliT	FliD chaperone	14	43
MotA	Force-generator; proton channel	32	43
MotB	Force-generator; spring	34	43

[a] Homologous to proteins in other species that serve as virulence factors.

TABLE A.4. Proteins involved in gene regulation.

Gene poduct	Function	Size (kd)	Gene map loc. (min)
FlgM	Anti-sigma factor	10	24
FlhC	Master regulator for middle genes	22	43
FlhD	Master regulator for middle genes	14	43
FliA	Sigma factor for late genes	28	43

Index